MASS

MASS

The quest to understand matter from
Greek atoms to quantum fields

JIM BAGGOTT

OXFORD
UNIVERSITY PRESS

OXFORD

UNIVERSITY PRESS

Great Clarendon Street, Oxford, OX2 6DP,
United Kingdom

Oxford University Press is a department of the University of Oxford.
It furthers the University's objective of excellence in research, scholarship,
and education by publishing worldwide. Oxford is a registered trade mark of
Oxford University Press in the UK and certain other countries

Published in the United States of America by Oxford University Press
198 Madison Avenue, New York, NY 10016, United States of America

British Library Cataloguing in Publication Data
Data available

Library of Congress Control Number: 2016960645

ISBN 978-0-19-875971-3

Printed in Great Britain by
Clays Ltd, St Ives plc

For Mike.
It's probably your fault…

CONTENTS

PREFACE

It has always been the dream of philosophers to have all matter built up from one fundamental kind of particle...

Paul Dirac[1]

It seems so simple.

You're sitting here, reading this book. Maybe it's a hardback copy, or a paperback, or an e-book on a tablet computer or e-reader. It doesn't matter. Whatever you're holding in your hands, we can be reasonably sure it's made of some kind of *stuff*: paper, card, plastic, perhaps containing tiny metal electronic things on printed circuit boards. Whatever it is, we call it *matter* or *material substance*. It has a characteristic property that we call *solidity*. It has *mass*.

But what *is* matter, exactly? We learn in school science class that matter is not continuous, but discrete. As a few of the philosophers of ancient Greece once speculated nearly two-and-a-half thousand years ago, matter comes in 'lumps'. If we dig around online we learn that we make paper by pressing together moist fibres derived from pulp. The pulp has an internal structure built from molecules (such as cellulose), and molecules are in turn constructed from atoms (carbon, oxygen, hydrogen). We further learn that atoms are mostly empty space, with a small, central nucleus of protons and neutrons orbited by electrons.

You might have also learned that protons and neutrons are not the last word on this subject. Particles thought to be the ultimate building blocks of matter or (more likely) whose internal structures are presently simply unknown are referred to by scientists as 'elementary'. According to this definition protons and neutrons are not elementary particles. They are composites, assembled from different kinds of quark, held together by gluons.

Okay, so things are a little more complicated than we might have supposed. But surely all we're really seeing here is successive generations of scientific discovery peeling away the layers of material substance. Paper, card, plastic; molecules; atoms; protons and neutrons; quarks and electrons. As we descend through each layer of matter we find smaller and smaller constituents. This is surely hardly surprising.

But then, just as surely, we can't keep doing this indefinitely. Just as the ancient Greek philosophers once speculated, we imagine that we should eventually run up against some kind of ultimately fundamental, indivisible type of stuff, the building blocks from which everything in the universe is made.

And it doesn't seem to require a particularly bold leap of imagination to suppose that, whatever it might be, there can be only *one* fundamental type of stuff. Or, at least, one fundamental type of stuff would seem simpler, or neater. The rest—electric charge, something called colour charge, flavour, spin, and many other things besides—would then just be 'dressing'.

In 1930, the English physicist Paul Dirac called this 'the dream of philosophers'. These were simpler times. The neutron hadn't yet been discovered (it was discovered by James Chadwick in 1932) and, so far as physicists of the time understood, all matter was composed of just two kinds of elementary particle—positively charged protons and negatively charged electrons. For a time, Dirac thought he had found a way to reconcile these, and the quote that I used to open this Preface continues: 'There are, however, reasons for believing that the electron and proton are really not independent, but are just two manifestations of one elementary kind of particle.'

Alas, Dirac was wrong. What he had stumbled across in the mathematical equations of his new quantum theory of the electron was not, after all, a fundamental relationship between the proton and the electron. He had deduced the existence of an altogether different kind of matter, which become known as antimatter. The positively charged entity that his theory predicted was not the proton. It was the anti-electron (or positron), discovered in studies of cosmic rays just a couple of years later.

After 1930 things just went from bad to worse. The dream became something of a nightmare. Instead of two elementary particles that might somehow be related, physicists were confronted by a veritable 'zoo' of different kinds of particles, many with seemingly absurd properties. It is a simple truth that modern science has undermined *all* our comfortable preconceptions about the physical universe, and especially the nature of material substance.

What we have discovered is that the foundations of our universe are not as solid or as certain and dependable as we might have once imagined. They are instead built from ghosts and phantoms, of a peculiar quantum kind. And, at some point on this exciting journey of discovery, we lost our grip on the reassuringly familiar concept of mass, the ubiquitous *m* that appears in all the equations of physics, chemistry, and biology.

To the ancient Greek atomists, atoms had to possess *weight*. To Isaac Newton, mass was simply *quantitas materiae*, the amount or quantity of matter an object contains. On the surface, there seem no grounds for arguing with these perfectly logical conclusions. Mass is surely an 'everyday' property, and hardly mysterious. When we stand on the bathroom scales in the morning, or lift heavy weights in the gym, or stumble against an immovable object, we pay our respects to Newton's classical conception of mass.

But when a single electron passes like a phantom at once through two closely spaced holes or slits, to be recorded as a single spot on a far detector, what happens to the mass of this supposedly 'indivisible' elementary particle in between? Einstein's most celebrated equation, $E = mc^2$, is utterly familiar, but what does it really mean for mass and energy to be equivalent and interchangeable?

The so-called 'standard model' of particle physics is the most successful theoretical description of elementary particles and forces ever devised. In this model, particles are replaced by quantum fields. Now, how can a quantum *field* that is distributed through space and time have mass, and what *is* a quantum field anyway? What does it really mean to say that elementary particles gain their mass through interactions with the recently discovered Higgs field? If we add up the masses of the three quarks that are believed to form a proton, we get only one per cent of the proton mass. So, where's the rest of it?

And then we learn from the standard model of inflationary big bang cosmology that this stuff that we tend to get rather obsessed about—so-called 'baryonic' matter formed from protons and neutrons—accounts for less than five per cent of the total mass-energy of the universe. About twenty-six per cent is dark matter, a ubiquitous but completely invisible and unknown form of matter that is responsible for shaping the large-scale structure of visible galaxies, galaxy clusters, and the voids in between. The rest (a mere sixty-nine per cent) is believed to be dark energy, the energy of 'empty' space, responsible for accelerating the expansion of spacetime.

How did this happen? How did the answers to our oh-so-simple questions become so complicated and so difficult to comprehend?

In *Mass*, I will try to explain how we come to find ourselves here, confronted by a very different understanding of the nature of matter, the origin of mass and its implications for our understanding of the material world.

One word of warning. The authors of works with pretentions to present popular interpretations of the conclusions of modern science tend to duck the difficult challenge of dealing with its mathematical complexity. There's the famous quote in Stephen Hawking's *A Brief History of Time*: 'Someone told me that each equation I included in the book would halve the sales.'[2] In previous books, I've tended to follow this rubric, limiting myself to a very small number of very familiar equations (see $E = mc^2$, above).

But the language of mathematics has proved to be enormously powerful in describing the laws of nature and the properties of matter. It's important to recognize that theorists will most often pursue a mathematical line of reasoning to see where it takes them, without worrying overmuch about how the mathematical terms that appear in their equations and the resulting conclusions should then be physically interpreted.

In the early years of the development of quantum mechanics, for example, the Austrian theorist Erwin Schrödinger bemoaned a general loss of what he called *anschaulichkeit*, of visualizability or perceptibility, as the mathematics became ever denser and more abstract. Theorists, supported by experiment or observation, may be able to prove that this mathematical equation represents some aspect of our physical reality. But there's absolutely no guarantee that we'll be able to interpret its concepts in a way that aids comprehension.

So, I've chosen in this book to reveal a little more of the mathematics than usual, simply so that interested readers can get some sense of what these concepts are, how physicists use them, and how they sometimes struggle to make sense of them. In doing this

I'm only going to scratch the surface, hopefully to give you enough pause for thought without getting too distracted by the detail.*

If you can't always follow the logic or don't understand the physical meaning of this or that symbol, please don't be too hard on yourself.

There's a good chance nobody else really understands it, either.

It's a real pleasure to acknowledge the efforts of Carlo Rovelli, who made some helpful and encouraging comments on the draft manuscript. Now I've never really expected family or friends to read my stuff, although it's always nice when they do (especially when they then say nice things about it). Obviously, I'm thankful to my mother for lots of things, but on this occasion I'm especially grateful, as she took it upon herself to read every word and provide helpful suggestions on how I might make these words simpler and more accessible. Now my mum has had no formal scientific education (she graduated with a degree in history from Warwick University in England when she was seventy-four), but she has boundless curiosity and enthusiasm for knowledge about the world. My hope is that if my mum can follow it...

I must also acknowledge my debts to Latha Menon, my editor at Oxford University Press, and to Jenny Nugee, who helped to turn my ramblings into a book that is hopefully coherent, no matter what it's made of.

Jim Baggott
October 2016

---

* Actually, I set myself the following constraints. No equations in the main text with more than two or at most three variables plus a constant ($E = mc^2$ has two variables, E and m, and one physical constant, c). There's a little more mathematical detail in the Endnotes for those interested enough to dig deeper.

LIST OF FIGURES

ABOUT THE AUTHOR

Jim Baggott is an award-winning science writer. A former academic scientist, he now works as an independent business consultant but maintains a broad interest in science, philosophy, and history, and continues to write on these subjects in his spare time. His previous books have been widely acclaimed and include:

Origins: The Scientific Story of Creation, Oxford University Press, 2015

Farewell to Reality: How Fairy-tale Physics Betrays the Search for Scientific Truth, Constable, London, 2013

Higgs: The Invention and Discovery of the 'God Particle', Oxford University Press, 2012

The Quantum Story: A History in 40 Moments, Oxford University Press, 2011

Atomic: The First War of Physics and the Secret History of the Atom Bomb 1939–49, Icon Books, London, 2009, re-issued 2015 (short-listed for the Duke of Westminster Medal for Military Literature, 2010)

A Beginner's Guide to Reality, Penguin, London, 2005

Beyond Measure: Modern Physics, Philosophy and the Meaning of Quantum Theory, Oxford University Press, 2004

Perfect Symmetry: The Accidental Discovery of Buckminsterfullerene, Oxford University Press, 1994

The Meaning of Quantum Theory: A Guide for Students of Chemistry and Physics, Oxford University Press, 1992

PART I

ATOM AND VOID

In which the concept of the atom is introduced by the philosophers of ancient Greece and evolves from indivisible, indestructible bits of matter to the atoms of chemical elements as we know them today.

1

THE QUIET CITADEL

This dread and darkness of the mind cannot be dispelled by
the sunbeams, the shining shafts of day, but only by an understanding of the outward form and inner workings of nature.

Lucretius[1]

I propose to start from some clear and simple beginnings, and
follow the breadcrumbs of observation, experiment, and logical reasoning into the heart of the mystery of matter. We'll begin
with the kinds of things we might deduce for ourselves just by
observing the world around us and pondering on its nature,
without the benefit of having access to a fully equipped physics
laboratory or a handy high-energy particle collider.

Now, I've managed to convince myself that this means starting
with the philosophers of ancient Greece. I say this not because
I think they can necessarily inform our understanding today—it
goes without saying that the ancient Greek philosophers didn't
have the benefit of a modern scientific education. All they could
do was apply a little logic and imagination to the things they
could perceive with their unaided senses, and this, I think, is a
great place to start.

We owe much of our common preconceptions of the nature
of matter to the physical world imagined by the ancient Greeks,

and especially those we refer to as the *atomists*. These were
Leucippus of Miletus (or Abdera or Elea, according to different
sources), who is thought to have lived in the middle of the fifth
century Before the Common Era (BCE); his pupil Democritus of
Abdera (born around 460 BCE); and their intellectual heir Epicurus
of Samos (born over a century later, around 341 BCE), who revived,
adapted, and incorporated this early version of the atomic theory
into a formal philosophy. In truth, our knowledge of precisely
what these philosophers said or how they structured their arguments is in some places rather vague. Epicurus argued that Leucippus
may not have even existed and that the credit for devising the
atomic theory belongs solely to Democritus. Only about 300
fragments of the writings attributed to Democritus have survived.
That may sound like a lot, but they pale in comparison with the
list of works compiled by the third-century CE biographer
Diogenes Laërtius, in his book *Lives of the Eminent Philosophers*.

According to Diogenes, Democritus wrote extensively on physics, cosmology, and mathematics, and on ethics and music. His
obsession with the human condition, and especially our sense of
happiness or cheerfulness, led him to become known as the 'laughing philosopher'. Much of what we know of Democritus' work
comes to us second-hand, from the commentaries of later philosophers, some of whom (such as Aristotle, born in 384 BCE) were
vocal, though seemingly respectful, *opponents* of the atomic theory.

The situation is a little better when it comes to the writings
of Epicurus. He penned several summaries of his work (called
epitomes), including one on his physical theory written to his
pupil Herodotus, and this, it would seem, is quoted in full by
Diogenes. The Epicurean philosophy was also the inspiration for
Roman poet and philosopher Titus Lucretius Caro's epic poem
De Rerum Natura (translated variously as 'On the Nature of Things'
or 'On the Nature of the Universe'), which was published around

55 BCE and which appears to be a relatively faithful adaptation of Epicurus' own thirty-seven-volume magnum opus, *On Nature*.*

It may yet be possible to learn more of Epicurus' particular brand of atomism from his own writings. A handsome villa in the Roman city of Herculaneum, thought to belong to Julius Caesar's father-in-law and sitting half-way up Mount Vesuvius, was buried in ash and debris in the eruption of 79 CE. Excavations in the eighteenth century uncovered an extensive library, containing more than 1,800 papyri (the 'Herculaneum papyri').[2] This is believed to be the personal library of the philosopher Philodemus of Gadara, born around 110 BCE. Philodemus studied in the Epicurean school in Athens, and many of the papyri contain key parts of *On Nature*, though these are badly damaged and many gaps remain.

Enough of the history. Let us now examine the logic. To be fair to these ancient thinkers, let's try temporarily to forget what we know about our modern world. We're going to indulge in a little armchair philosophizing, and we don't want to be distracted by the trappings of our modern existence. Let's imagine ourselves strolling barefoot along an Aegean beach in fifth-century BCE Western Thrace in Greece, about 17 kilometres northeast of the mouth of the Nestos River (see Figure 1). It's a fine day. The Sun is shining and a gentle breeze blows inland. As we stroll, we're preoccupied with a single question.

How is the world put together?

Before we can really begin to address this question we need to establish some ground rules. Living as we do in fifth-century BCE Greece, our lives and many of our daily rituals are governed by

---

* In the opening passage of Book II of *On the Nature of the Universe* (Penguin, London, first published 1951), Lucretius compares the joy of philosophical reflection to standing aloof in 'a quiet citadel, stoutly fortified by the teaching of the wise', gazing down on struggling humanity. I suspect this is the earliest depiction of an 'ivory tower' as you're likely to find.

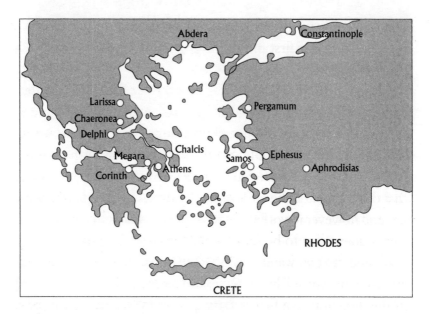

Figure 1. Greece and Western Asia Minor.

a need to pay our respects to the gods. Let's agree that, irrespective of our own personal beliefs and prejudices, we will seek an answer to our philosophical question that does not rely ultimately on some kind of divine intervention. As we look around us, we see blue sky, a sandy beach, a restless sea and, on a distant hill, a flock of sheep grazing on green grass. No gods.

If we deny that the gods have any role in creating and shaping the material world, then we eliminate the element of unpredictability, chance causation and the sense of 'anything goes' that might otherwise be identified with the 'will of the gods'. We then allow ourselves an opportunity to discern something of the world's underlying *natural* order.

Setting legend and superstition aside, it is our common experience that there are no miracles. Objects in the material world don't suddenly appear, from nowhere or from nothing. And, although they undoubtedly change over time, objects likewise don't suddenly vanish into thin air. This means that we can move on quickly to our first important logical deduction. *Nothing can come from nothing.*[3]

Let's go further. Just as we see no evidence for the gods intervening in the material world, so too we see no evidence for any influence of what we might call the human soul or spirit. Again, this doesn't mean to imply that the soul or spirit doesn't exist or isn't in some way connected to the way the mind works. It's just that these appear to be quite distinctly different things. Whatever it is and however it works, it is my experience that my mind (soul, spirit) does seem to be rather firmly fixed in my head or in my body, and can't go wandering off into the external material world, at least while I'm still living. What this means is that we're heading in the direction of a firmly *materialist* or *mechanistic* philosophy. Our external material world is shaped *only* by non-sentient, physical mechanism.

A moment's reflection as we continue our stroll leads us to conclude that there exists an astonishing variety of different *forms* of matter. Look around. We see rock, soil, sand, water, air, living creatures. Drawn up on the beach ahead of us is a small wooden fishing boat, abandoned for now by its owner. We can see why. It is holed just below the waterline. A crude attempt at repair has reduced some of the wooden hull to a fine sawdust, which has gathered on the sand beneath and is now catching in the breeze. No object in the material world can be created from nothing and, logic suggests, objects can never be reduced to nothing. Left to itself the sawdust will disperse, scattered to the winds. But, you now suggest, the matter that was once solid wood and is now sawdust has simply changed form, and although it will blow away it doesn't disappear. I tend to agree.

The sawdust shows that, as a result of some mechanical action, the solid wood of the boat's hull can be finely divided. But then, what if we could divide the sawdust even more finely? And then divide it some more? Couldn't we keep on doing this, endlessly dividing the matter into smaller and smaller pieces, *ad infinitum*? Wouldn't we end up dividing it completely into nothing, and so contradicting our earlier conclusion?

This reminds you of a famous paradox devised by one of our contemporaries, Zeno of Elea. The one about the race between the Greek hero Achilles and the tortoise. I've heard it before, but you tell the story well and it is worth repeating. It is clear that Achilles and the tortoise are unevenly matched. But Achilles has a strong sense of honour and fair play, and he agrees to give the tortoise a head start, no doubt confident of ultimate victory. So, Achilles waits until the tortoise has reached a certain position—half-way to the finish line—before setting off. But by the time Achilles has reached half-way, the tortoise will inevitably have moved on a certain additional distance. By the time Achilles has reached that additional distance, the tortoise will have moved on a little further. We can go on like this for ever, it seems. Each time Achilles reaches the point where the tortoise was, the tortoise has moved a little further ahead. It seems that Achilles will never overtake the tortoise.

At the heart of Zeno's paradox lies the seemingly innocent observation that a line can be divided into an infinite number of distinct points. But if there's an infinity of points between start and finish, then no kind of motion through each and every point can be conceived that will allow us to get from start to finish in a *finite* time. Zeno is a pupil of Parmenides of Elea, and philosophers of the Eleatic school argue that, contrary to appearances, all change is an illusion. There is no motion because motion is simply impossible. This, according to Parmenides, is the 'way of truth'. Appearances, in contrast, are deceptive and so cannot be trusted. He calls this the 'way of opinion'. Hmm...

We ponder on this for a while. We debate some more and agree that it's really rather illogical bordering on absurd to deny ourselves access to everything we can learn about the material world by engaging our senses. Why not trust them? Why not rely on how things appear? But then how do we resolve Zeno's paradox?

You're struck by a thought. You explain that the paradox is actually based on a confusion. Whilst it is correct to suggest that a continuous line can be *mathematically* divided into an infinity of points, this does not mean that a distance or an area or a volume in the real world can be *physically* so divided. What if the material world is *not* continuous and endlessly divisible, but is instead composed of discrete, indivisible, or uncuttable parts? You use the Greek word *atomon*, or *a-tomon*, meaning an entity that cannot be cut or divided.

It's an intriguing line of argument, and it leads us to another conclusion. Matter cannot be divided endlessly into nothing: it can be divided only into its constituent atoms.[4]

I sense some problems, however. We perceive change in the external material world because matter changes over time and changes from one form into another. (Think about the lake that freezes over with ice in the winter.) But underpinning all the different kinds of matter are indestructible atoms, right? Ah, but if the atoms are indestructible and unchangeable, and therefore eternal, just *how* can they be responsible for the change that we perceive?

You ponder this question for a while, then suddenly snap your fingers. Change happens because the atoms are *constantly moving*, colliding with each other and forming different associations with each other that represent the different forms that matter can take. Okay, I can live with that. But what, can I ask, are these atoms supposed to be moving *in*? I sense you're on a roll. You come right back and announce with conviction that the *solid atoms are moving in empty space*, also known as the 'void'.⁵ This kind of thing will doubtless sow some seeds for future philosophical argument. (Aristotle was firmly against the idea of the void and is credited with the declaration that 'nature abhors a vacuum'.) But we'll run with it for now.

So, just by looking at the world around us and thinking logically about its construction and the nature of change we've come to the conclusion that all matter exists in the form of atoms moving restlessly in space. With a little more intellectual effort we can sharpen this basic description to fit a few more of our observations.

We can suppose that by mixing different proportions of 'hard' atoms and empty space we can construct the wonderful variety of material substance in all its forms, which the Ancient Greeks reduced to four basic 'elements'—earth, air, fire, and water.⁶ Although he did not acknowledge any influence of Leucippus or Democritus on his work, the great philosopher Plato (born around 428 BCE*) developed an elaborate atomic theory. He represented each of the four elements by a geometrical (or so-called 'Platonic') solid, and argued in the *Timaeus* that the faces of each solid could be further decomposed into systems of triangles, representing the elements' constituent atoms. Rearrange the patterns of triangles—rearrange the atoms—and it is possible to convert one element into another and combine elements to produce new forms.⁷

Plato fixed on triangles, but the early atomists (and, subsequently, Epicurus) argued that atoms must possess different *shapes*, some more rounded, with gentle curves, some more angular, sharp-edged, and 'spiky', with barbs and hooks. As the atoms collide they stick together to form composites (I guess what we would today call molecules). The different textures of the resulting

---

* Or 427 or four years later.

combinations are ultimately responsible for the properties and behaviours of the material substances thus formed.

The release of films of atoms from these substances cause us to have sensory perceptions. We perceive colour through the 'turning' or changing positions of the atoms that enter our eyes, the texture of the atoms on our tongues cause taste sensations, and so on. The atomists did not imagine the atoms in such combinations to be held together by any kind of force, but rather by the interlinking of their shapes. For example, Lucretius suggested that that bitter taste of seawater could be attributed to the presence of 'rough' atoms. These rough atoms can be filtered out by passing seawater through layers of earth (as they have a tendency to 'stick' to earth), allowing the 'smooth' atoms through and so giving a much more palatable fluid.[8]

Democritus suggested that there was no limit to the number of different possible shapes that atoms could possess, and that they could, in principle, be of any size. Epicurus was more circumspect. He argued that there must be a limit to the number of shapes. And atoms are small beyond the limits of perception—if we can see it then it's not of atomic size.

This is all very fine, but if the atoms are supposed to be perpetually moving, what is it that is *making* them do so? Aristotle (who was a student of Plato) didn't buy it.[9] The early atomists never really explained what causes such motion, even though this is absolutely essential to the theory.

Epicurus provided something of an answer by crediting the atoms with *weight* which causes a 'downward' motion through the infinite cosmos, as can be observed in the behaviour of any and all substantial things on Earth. The atoms are kept in motion either by their own weight or by the impacts of collisions with other atoms.[10] But anyone who has ever been caught in a heavy downpour will have noticed how raindrops can sometimes appear to fall vertically. If the atoms are not subject to any other force except for a kind of gravity that pulls them downward, then why wouldn't they simply fall straight down, so avoiding chance collisions? According to later commentators, Epicurus admitted that they sometimes 'swerve':

> When the atoms are travelling straight down through empty space by their own weight, at quite indeterminate times and places they swerve ever so little from their course.[11]

This argument doesn't hold up well under scrutiny, and in fairness I should point out that although we've no reason to doubt later sources, no such comment can be found in the surviving writings of Epicurus himself.

If we accept that atoms are endlessly in motion, how do we reconcile this with the fact that large, observable objects are still, or move only slowly? The atomists argued that we don't see the motion because we simply can't. Remember that flock of sheep grazing on the distant hillside as we strolled along the beach? It serves as an example. From this distance, we can't discern the movements of individual sheep. We see only a vague—and seemingly stationary—white blur on the green hillside.[12]

One last challenge. Why should we believe in the reality of atoms if they are so small that we can never see them? Isn't this just the same as believing in gods or any other construction of the imagination that we could invoke to explain observed phenomena, but for which we can gather no evidence using our senses? The atomists advise us to stick to our mechanistic instincts. Although we can't see these invisible entities, there is plenty of visible evidence which alerts us to their existence. There are effects for which atoms must be the cause.

Just look at many natural phenomena, such as wind, odour, humidity, or evaporation. We're all too familiar with such things, although we cannot see the agencies that cause them. Likewise, we're aware of the imperceptible wearing down of a ring worn on a finger, or a ploughshare in a furrow, or the cobblestones under your feet, or the right hands of bronze statues at the city gates worn thin by the greeting touch of travellers, although we cannot see the particles that are lost through such slow decay. Nature must work through the agency of invisible atoms.[13]

Perhaps there's even more direct evidence. Imagine a quiet space, inside an ancient building. A high window admits a beam of sunlight which lightens the darkness. As we look more closely we notice a multitude of tiny particles dancing in the beam. What's causing this dance of the dust motes? Isn't this evidence of invisible atomic motions?[14]

It's hard to fault this logic although, of course, this is not the right conclusion. Dust motes dance in a sunbeam because they are caught in currents of air, not because they are buffeted by the chaotic motions of atoms. It's just a question of *scale*. But this description would be perfectly appropriate when applied to fine grains of pollen suspended in a fluid, whose random, microscopic motions were observed and reported by Scottish botanist Robert Brown in 1827 and which is now collectively called *Brownian motion*. In 1905, a young 'technical expert, third class' working at the Swiss patent office in Bern published a paper explaining how Brownian motion is visible evidence for the random movements of the invisible atoms or molecules of the fluid. His name was Albert Einstein (and we'll get to him soon enough).

So, according to the ancient atomists matter is composed of atoms moving restlessly in the void. Different forms of matter are constructed from different mixtures of atoms and void, and from different combinations of atoms. Changes in these combinations and mixtures cause matter to change from one form to another, and films of atoms released from material objects cause sensory perceptions. The atoms possess properties of size, shape, position, and weight, and sometimes 'swerve' into each other as they fall, but are so small as to be invisible to our eyes.

This is all perfectly logical and wonderfully well argued. But there remains a problem with this structure that threatens to undermine it completely. We've arrived at these conclusions by trusting that our perceptions of the external world provide us with a faithful representation, on which we can apply our logic with confidence. However, the atomists agreed that although these perceptions are in some way *caused* by atoms that exist in reality, the sensations that result—colour, taste, odour, and so on—are not 'real' in this same sense. The atoms themselves have no sensory properties—for example, atoms cause the sensation of colour or a bitter taste but they are not in themselves coloured, or bitter. The sensations we experience are constructs of our minds and exist only in our minds. There will be a lot more on this topic in Chapter 2.

But surely everything we know or can deduce is shaped by these very perceptions. If we accept that they exist only in the internal workings of our minds, then it seems we are denied access to the external reality that we're trying so hard to create a structure for. It seems that on this point the laughing philosopher was really rather pessimistic: 'We know nothing in reality;' he declared, 'for truth lies in an abyss.'[15]

Five things we learned

1. Matter is 'substantial'. It cannot suddenly appear from nothing and it cannot be divided endlessly into nothing.

2. Therefore all matter must consist of ultimate, indivisible components which we call atoms.

3. Atoms move endlessly in empty space—the void. Different forms of matter are composed of mixtures of atoms with different shapes and different proportions of hard atoms and void.

4. Atoms move because they possess weight. As they fall they sometimes 'swerve' and collide with each other.

5. Atoms possess the properties of size, shape, position (in the void), and weight. They also cause sensory perceptions in our minds, of colour, taste, odour, and so on, although the atoms themselves do not possess these properties directly.

2

THINGS-IN-
THEMSELVES

Though we cannot know these objects as things in themselves,
we must yet be in a position to at least think [of] them as things
in themselves; otherwise we would be landed in the absurd
conclusion that there can be appearance without anything that
appears.

Immanuel Kant[1]

I want to reassure you that this business about perception and
the nature of external reality which so exercised Democritus is
not some philosophical mire of our own making. It is not something from which we can extricate ourselves only by engaging in
seemingly interminable nit picking. It is a fundamental problem
that will have a profound impact on our understanding of matter
and it will be a recurring theme. I fear that anyone suffering the
delusion that science is free from this kind of philosophical
wrangling is most likely in for bit of a shock.

I propose in this chapter to move on quite quickly from the
ancient Greek atomists to some of the great philosophers of the
seventeenth and eighteenth centuries. Now, in doing so I don't
want to give you the impression that nothing of consequence
was discussed, debated, or written by philosophers in the sixteen
centuries that passed in between. But I think it's fair to say that
much of the attention of Western philosophers during this period

was absorbed by the challenge of reconciling the philosophies of ancient Greece and Rome with the theologies of the 'Abrahamic' religions of Christianity, Judaism, and Islam.[2]

Some of the tenets of ancient Greek philosophy were preserved, even as the Roman Empire began its slow decline, initially by a few scholars with some facility in the Greek language. Not all of these scholars were wholly sympathetic, however. For example, the second-century CE Christian philosopher Quintus Septimius Florens Tertullianus (Tertullian, sometimes referred to as the 'father of Western theology'), despised Greek philosophy, declaring the Greeks to be the 'patriarchs of heretics'.[3]

The Greeks' pronouncements on the nature of the soul, resurrection, and the creation simply didn't fit the demands of a theology based on the notion of an all-powerful, omniscient, omnipresent, Christian God. It was perhaps inevitable that philosophical inquiry into the nature of the material world would get dragged into the debate. 'Natural philosophy' became inextricably tangled with theological questions, such that the lines we draw today between these disciplines became greatly blurred, or even non-existent.

But the emphasis on scholarship slowly returned, first through schools established by monasteries and cathedrals, some few of which eventually developed in the twelfth and thirteenth centuries into universities. The rise of academic philosophy and theology helped to renew interest in the ancient Greeks, even though formal teaching of some of their works was generally forbidden (e.g., the teaching of Aristotle's metaphysics and natural science was prohibited by the statutes of the University of Paris in 1215).

This burgeoning interest in a selection of the ancient Greek texts proliferated new translations. By the middle of the thirteenth century the climate had changed sufficiently to allow the Italian Catholic priest and theologian Thomas Aquinas to set about the task of rehabilitating Aristotle. The resulting 'Thomist' philosophy is Aristotle blended with many other sources, ancient and medieval. This was only a partial or selective rehabilitation: Aquinas served two spells as regent master of theology at the University of Paris, so Thomism is really a theology or a philosophy with a distinctly Christian gloss. In this way, Aristotle's pronouncements on the nature of matter, his Earth-centred cosmology based on perfect circular motion, caused by a prime mover, became enshrined in Christian religious orthodoxy.

So, what *did* Aristotle have to say about the nature of material substance? He had struggled to reconcile Democritus' rather passive, unchanging atoms with the observation of a very lively, actively changing material reality. And, as we have seen, he rejected completely the notion of the void. The atomists' rebuttal of Parmenides and Zeno arguably required that space and time should also be conceived to be 'atomic' in nature, with ultimate limits on how finely units of space and units of time can be divided.* Aristotle preferred to think of space and time as continuous, and any object taking up room in a continuous three-dimensional space is in principle infinitely divisible, thereby making atoms impossible. But he was nevertheless still sympathetic to the idea that a substance could be reduced to some kind of smallest constituent. It was just that the atomists had gone too far. They had been *too* reductionist in their approach.

His solution was an alternative structure based on the idea of *natural minima*, meaning the smallest parts into which a substance can be divided without losing its essential character. Now, natural minima are in principle *not* atoms, at least in the sense that the Greek atomists understood the term. Natural minima can in principle be divided. It's just that when they are divided beyond a

---

* This idea is firmly back with us today—see the Epilogue.

certain limit they no longer represent or are characteristic of the original substance. The solid wood of the holed fishing boat is divided into particles of sawdust, but the sawdust is still wood. Keep dividing the sawdust further and we eventually cross a threshold to something that is no longer wood.

Aristotle also endowed naturally occurring objects with *form*, a theory adapted from the teachings of his mentor, Plato. A tree consists of natural minima *and* possesses the form of 'tree-ness'. It is its form that makes an object what it is and governs its properties and behaviour. In the interpretation fashioned by Aquinas, this became a *substantial form*, which is not reducible to component parts. A tree is a tree, and cannot be reduced to component parts, such as trunk, branches, leaves, and so on, as the components are not 'trees'. Cut the tree down and use the wood to build a fishing boat and the substantial form of 'tree-ness' is lost.

This construction was a gift to medieval theologians. Humans can build boats, but only God can make a tree. The fusion of matter and form to make objects is readily applied also to the fusion of body and soul. It was a handy explanation for *transubstantiation*, the bread and wine used in the sacrament of the Eucharist carrying the substantial forms of the body and blood of Jesus Christ.

It would be all-too-easy to view this as a period of intellectual activity that was in some way stifled or imprisoned by some Machiavellian creatures of the established Church. It's certainly true that whilst philosophers were generally free to think what they liked, they were not free to write and disseminate what they thought without fear of accusations of heresy and all that this entailed. But the seeds of slow and quiet intellectual reformation were actually sown by leading Church figures, including several fifteenth- and sixteenth-century Popes: Pius II, Sixtus IV, and Leo X.

These were *Renaissance humanists*, who helped to establish *studia humanitatis*, what we know today as 'the humanities', a rediscovery of the learning and values of ancient Greece and Rome. The study of the humanities became possible in ways that were very different from the narrow prescriptions of medieval scholarship. The influence of Renaissance humanism can't be over-stated. It was to become one of the most important forces for change in human intellectual history.

In January 1417, the Italian scholar and manuscript-hunter Gian Francesco Poggio Bracciolini discovered a copy of Lucretius' *De Rerum Natura* languishing in a German monastery. He sent a copy to his friend Niccolò de' Niccoli, known to us today for inventing cursive, or italic, script. De' Niccoli agreed to transcribe it, though he appears to have sat on it for at least twelve years. Eventually, more copies were made (more than fifty copies from the fifteenth century have survived to the present day), and Johannes Gutenberg's ingenious invention helped spread Lucretius' poem across Europe.

But it was not what Lucretius had to say about the atomic theory of Epicurus that fired imaginations in this period. More interesting to fifteenth-century readers was what he had to say about the 'natural order'; the basis for philosophizing without the constant intervention of one (or more) deities, a natural order which appears not to have been designed specifically with humans in mind. Lucretius also talked about the death of the soul and the fallacy of the afterlife, and of the cruelty and superstitious delusions of organized religion.*

Safe to say, this did not play well. Printed editions of Lucretius' poem appeared with warnings and disclaimers, and in the

---

* Lucretius also identified the 'swerve' in atomic motions with the notion of free will. Consequently, Harvard professor of the humanities Stephen Greenblatt chose to title his 2012 Pulitzer prize-winning telling of the story of Bracciolini's discovery and its aftermath *The Swerve: How the Renaissance Began* (Vintage, London, 2012).

early sixteenth century it was banned from Italian schools. The Dominican friar Giordano Bruno was a vocal advocate of the Epicurean philosophy and of a cosmology that placed the Sun—not the Earth—at the centre of the universe, as proposed by Nicolas Copernicus in his book *De Revolutionibus Orbium Coelestium* ('On the Revolutions of the Heavenly Spheres'), published in 1543. On his ill-advised return to Italy from wandering around Europe, Bruno was arrested in May 1592 and imprisoned by the Inquisition. He refused to recant and was burned to death in February 1600.

It was clear that philosophers still needed to proceed with great caution. But by the seventeenth century they had sufficient intellectual freedom to establish structures for interpreting and understanding the physical world that were once again largely materialistic or mechanistic in nature. This was a world that might have been designed by God but whose mechanisms appeared free of overtly divine intervention. And they also gained the freedom to begin the slow process of disentangling philosophy from theology.

Again, it is probably a mistake to think of this as a triumph of science or rational thinking over religious superstition. Many of the seventeenth-century philosophers who ushered in this new 'age of reason' were motivated by efforts to understand the world that God had designed and created and to reconcile their conclusions with Christian doctrine. These were still philosophers, forming two broad and overlapping groups that we tend to think of today as 'mechanical' and 'classical modern'.

Among the mechanical philosophers, we would count Francis Bacon (born 1561), Galileo Galilei (1564), Johannes Kepler (1571), Pierre Gassendi (1592), Robert Boyle (1627), Christian Huygens (1629), and Isaac Newton (1642). Among the classical modern philosophers, we would count René Descartes (1596), John Locke

(1632), Baruch Spinoza (1632), Gottfried Leibniz (1646), George Berkeley (1685), and, stretching into the eighteenth century, David Hume (1711) and Immanuel Kant (1724). Today we tend to think of the former group as 'scientists' (or at least heralds of the scientific revolution) and the latter as 'philosophers'. In truth they form a near-continuous spectrum, varying only in the nature and methods of their inquiry. Many of the 'scientists' engaged in philosophical (and theological) reflection and many of the 'philosophers' engaged in experiment, or at least acknowledged the conclusions of experimental science. Descartes was also a mechanical philosopher.*

Boyle was a noted experimentalist, conducting studies in medicine, mechanics, hydrodynamics, and the properties of gases.† He also dabbled in alchemy and was an ardent Christian, advocating the 'argument from design' for the existence of God, encouraging the translation of the Bible and the dissemination of its messages.[4] He was an admirer of James Ussher, the Archbishop of Armagh, whose literal analysis of the Book of Genesis had led him to conclude that the Earth had been created by God in 4004 BCE, on 22 October, around 6 pm.

Of the early mechanical philosophers, Gassendi and Boyle were perhaps the most influential in reintroducing the idea of atoms (Descartes couldn't get past his dislike of the void). Gassendi embarked on an ambitious attempt to reconcile the Epicurean philosophy with Christianity, and his atoms were

---

* My good friend Massimo Pigliucci, a scientist-turned-philosopher working at the City University of New York, suggests that history regards Descartes as a 'philosopher' only because he got the physics wrong.

† Influenced by Francis Bacon and his laboratory assistant Robert Hooke, Boyle developed one of the first philosophies of experiment, which (perhaps inevitably) lends priority to observation and experiment, from which theory should then be inducted or deduced.

recognizably those of Epicurus. But Boyle's atoms were rather different.

In *The Scientist's Atom and the Philosopher's Stone*, first published in 2009, contemporary philosopher of science Alan Chalmers carefully traces the evolution of the sanctioned, orthodox Aristotelian description of matter into the more recognizable atomism adopted by the mechanical philosophers at the beginning of the scientific revolution. A key figure in this evolution was Paul of Taranto, a thirteenth-century Franciscan monk, who is believed to have written treatises on alchemy under the penname of Geber, 'borrowing' the name from a famous tenth-century Muslim alchemist Abu Mūsā Jābir ibn Hayyān (in the West 'Jābir' was Latinized as 'Geber'). The writings of 'pseudo-Geber', together with the medieval adaptation of Aristotle's natural minima and the atomic theory of Democritus, were powerful influences on Daniel Sennert, a renowned seventeenth-century German physician and professor of medicine at the University of Wittenburg.

Sennert also dabbled in alchemy. In one famous experiment he reacted metallic silver with aqua fortis (nitric acid) to produce a new compound which is not simply a mixture of the initial ingredients (we would today identify the product of this chemical reaction as silver nitrate). The resulting solution could be filtered without leaving the residue that would be expected if particles of silver had simply become dissolved in the acid. Adding salt of tartar (potassium carbonate) to this solution caused another compound (silver carbonate) to precipitate. Filtering, washing, and heating this precipitate would then recover the metallic silver, in its 'pristine state'.

This kind of experiment causes all sorts of problems for Aristotle's theory of forms. It is clear that the natural minima of silver persist unchanged through all these chemical transformations, since they are present at the start and can be recovered at the end. Sennert concluded that the natural minima of silver serve as *components* in the natural minima of each of the compounds formed through the sequences of chemical transformations. But these compounds are substances in their own right, with properties that are quite distinct from those of the starting materials. He was therefore obliged to admit the possibility of a *hierarchy* of forms. In other words, the forms themselves can be transformed and new forms created.

Although he did not openly acknowledge it in his writings, Boyle's views owed much to Sennert. He dismissed as unnecessary all reference to the Aristotelian forms, and argued that the properties and behaviour of a material substance can be traced to the natural minima from which it is composed. Boyle rarely used the word 'atom', although this is undoubtedly where he was heading (and to simplify the following discussion I'll revert to 'atoms' in preference to natural minima).

Boyle accepted that the atoms are too small to be perceived but assumed them to possess size (and hence weight), shape, and motion. They are physically indivisible or at least, like Sennert's silver atoms, they remain intact through chemical and physical transformations. This looks very much like the atomism of Democritus and Epicurus, although Boyle had arrived at this structure in his own way and on his own terms. Boyle's atomism was not a 'rediscovery' of the ancient theory.

The mechanical philosophers drew on the talents of a new generation of artisan instrument-makers, developing the technology needed for systematic observation or experimentation, such as telescopes, microscopes, clocks, and vacuum pumps. Boyle's experiments with an air pump (built by his laboratory assistant at Oxford, Robert Hooke) led him to deduce the relationships between gas pressure and volume that are the basis for

Boyle's law. Squeeze a gas into a smaller volume whilst maintaining a constant temperature and the pressure of the gas increases in direct proportion. Allow a gas to expand into a larger volume whilst maintaining a constant temperature and the pressure of the gas falls in direct proportion. If the temperature is fixed, pressure multiplied by volume is constant.

But these kinds of experiments were not so refined that they resolved the problems that had confronted the ancient Greeks. Just how *do* you draw conclusions about atoms that by definition cannot be observed directly in the properties and behaviour of matter on a large scale?

Boyle's solution was to suggest that properties and behaviour observed to occur 'universally' in matter of all kinds can logically be ascribed to the atoms themselves. In other words, what we see in our large-scale, macroscopic world of experience applies equally to the microscopic entities from which substance is composed. This is arguably no more advanced than Lucretius, who suggested we maintain our mechanical sensibilities when observing natural phenomena. But the microscopic atoms are still required to possess properties (such as indivisibility) that are simply absent from our macroscopic world, and so these are properties for which the new experimentalists could establish no evidence.

As the mechanical philosophers fussed with their instruments, the classical modern philosophers set about the task of defining what, if anything, could be learned about the material world from reason alone. Descartes set out to build a new philosophical tradition in which there could be no doubting the absolute truth of its conclusions. From absolute truth, he argued, we obtain certain knowledge. However, to get at absolute truth, Descartes: '...thought I ought to...reject as being absolutely false everything in which I could suppose the slightest reason for doubt'.[5] This meant rejecting all the information about the world that he received using his senses.

He felt he could be certain of at least one thing: that he is a being with a mind that has thoughts. He argued that it would seem contradictory to hold the view that, as a thinking being, he does not exist. Therefore, his existence was also something about which he could be certain. *Cogito ergo sum*, he famously concluded: 'I think therefore I am.'

Having proved his own existence (at least to himself), Descartes went on to present a number of proofs for the existence of God, conceived as an all-perfect being which, he established by logical reasoning, must be a force for good. Now, if Descartes' ideas of objects existing in the world are deceptions, then this would mean that God was deceiving him or allowing him to be deceived. If we take deception to be a source of imperfection, then this would seem to contradict the idea of God as an all-perfect being. Descartes concluded that his ideas of physical objects must therefore be the direct result of the existence of these objects in the external world.

But, just as the ancient Greeks had done, he recognized the distinction between the objects themselves and the sensory perceptions they create, and that the latter do not necessarily convey an accurate representation of the objects 'in reality'. We perceive different colours, smells, tastes, sounds, and sensations such as heat, hardness, and roughness. The sources of all these different perceptions would seem to be *created* by the objects themselves, but this does not mean that they are intrinsic *properties* of the objects.[6]

This distinction was greatly sharpened by the English philosopher John Locke in 1689. He argued that, no matter how finely we might divide a substance, what we get retains certain intrinsic, or *primary*, qualities, such as shape, solidity, extension (in space), and motion. If the ultimate constituents of matter are atoms, then we can expect that these atoms will possess the primary qualities.

But there are also *secondary* qualities that are not intrinsic but which result from interactions with our sensory apparatus: eyes, ears, nose, tongue, and skin. Individual atoms might possess shape and solidity and so on, but they do *not* possess colour, sound, or taste. Colour, for example, is the result of an (unspecified) interaction between the atoms and our eyes.[7]

This seems to make sense, but philosophers like nothing better than a good argument, and they were far from done. Try as he might, the Irish philosopher George Berkeley couldn't see the difference between primary and secondary qualities. Whilst it might be possible to distinguish between them, as far as he was concerned it is practically impossible to *separate* them. If we can't conceive of an object possessing shape and solidity without at the same time possessing colour or sound, then this suggests that, despite the distinction, primary and secondary qualities have much the same *status*. Berkeley was happy to concede that secondary qualities exist only in our minds. What he was saying is that primary qualities exist only in our minds, too.[8]

The Scottish philosopher David Hume was inclined to agree. He concluded simply that we have no means of knowing what, if anything, exists beyond our capacity to experience it through perception and so the question is, broadly speaking, meaningless. His strategy was to consign all speculation about the nature of reality to metaphysics (meaning literally 'beyond physics') and to adopt a fairly negative attitude to any claim to knowledge achieved through metaphysical reasoning and speculation. In arguing this position Hume helped to establish a philosophical tradition known as *empiricism*.

In an empiricist philosophy, knowledge of the world that is not gained through direct experience is rejected as meaningless metaphysics. This does not necessarily mean that there is no such thing as a reality existing independently of perception: the Moon is still there, even if nobody looks. But it does mean that we might have to manage our expectations. At best, we gain knowledge of an *empirical reality*—the reality manifested as effects that we can directly perceive, or make measurements on.

The great German philosopher Immanuel Kant was deeply influenced by Hume's work, but he came to deny Hume's conclusion that it is therefore impossible to acquire knowledge through anything other than experience. He differentiated between what he called *noumena*, the objects or things-in-themselves, and *phenomena*, the things-as-they-appear as we perceive and experience them.

Hume would likely have argued that noumena are metaphysical and meaningless, but Kant argued that, through phenomena, the noumena impress themselves upon our minds because our minds have what he called *sensible intuitions*, specifically of space and time. We construct space and time in our minds in order to make sense of the things we perceive in the external world around us.

Kant claimed that it made no sense to deny the existence of things-in-themselves, as there must be some things that cause appearances in the form of sensory perceptions (there can be no appearances without anything that appears). But he agreed with Hume that whilst the things-in-themselves must exist, we can in principle gain no knowledge of them.

What does this kind of logic mean for atoms? It can be argued that these entities produce appearances both obvious (as Lucretius had noted) and more subtle (as the alchemists and early mechanical philosophers had deduced). But it is a simple and rather stubborn fact that they cannot be perceived. Some later scientists and philosophers who would declare themselves to be staunch empiricists, such as Austrian physicist Ernst Mach, would come to dismiss atoms as metaphysics and deny that they really exist. It was certainly true that, for as long as experimental science was unable to access phenomena that could be attributed more directly to atoms, they would remain, at best, a 'working hypothesis'.

This was nevertheless good enough for many of the practically minded mechanical philosophers, who had long before embarked on a voyage of empirical discovery. They tended to their instruments and conducted careful, systematic experiments, varying first this, then that. And there was something else. They discovered that the laws of nature are written in the language of mathematics.

Five things we learned

1. The sixteenth- and seventeenth-century mechanical philosophers disentangled themselves from Aristotle's description of nature and returned to an atomic theory not much different from that of the Greek atomists.
2. The classical modern philosophers, meanwhile, debated the extent to which we can gain knowledge from observation and reasoning alone.
3. Locke distinguished between an object's primary qualities (shape, solidity, extension in space, and motion) and its secondary qualities (colour, taste, odour, etc.). Secondary qualities do not 'belong' to an object: they exist only in our minds.
4. Berkeley (and later Hume) argued that as everything we learn about nature must be derived via our senses, then primary qualities must exist only in our minds, too.

5. Kant argued that it makes little sense to believe that that the things we perceive do not exist in reality. He distinguished between noumena (things-in-themselves) and phenomena (things-as-they-appear), but admitted that although the former must exist, we can only gain knowledge of the latter.

3

AN IMPRESSION OF FORCE

The alteration of motion is ever proportional to the motive
force impressed; and is made in the direction of the right line
in which that force is impressed.

Isaac Newton[1]

Despite their somewhat dubious status, atoms (or 'corpus-
cles') of one kind or another were fairly ubiquitous in the
physical descriptions adopted by the seventeenth-century
mechanical philosophers.[2] Like Boyle, Isaac Newton remained
relatively unperturbed by the fact that it was impossible to gain
any direct evidence for their existence. The first edition of
Newton's *Mathematical Principles of Natural Philosophy* was published in 1687, and in a second edition published in 1713, he added
four 'rules of reasoning'. Rule III, which concerns the 'qualities of
bodies', contains a partial summary of his atomism. In it he
insists that an object's shape, hardness, impenetrability, capabil-
ity of motion, and inertia all derive from these same properties
manifested in the object's 'least parts'.[3]

Like Boyle, Newton was happy to accept that the visible properties and behaviour of a macroscopic object (such as a stone)
can be understood in terms of precisely the same kinds of properties and behaviour ascribed to the invisible microscopic atoms

from which it is composed. The message is reasonably clear. Let's not worry overmuch about the status of metaphysical entities that we can't see and can't derive any direct evidence for. Let us instead devote our intellectual energies to the task of describing the motions of material objects that we *can* observe and perform measurements on (a discipline called kinematics). Then let's further seek underlying *explanations* for this motion (a discipline called dynamics). We do this reasonably secure in the knowledge that anything we can discover about such macroscopic objects will likely apply equally to their constituent microscopic atoms.

Where do we start? By acknowledging that any observations or measurements of the motions of objects require a framework in which these can be made. This is a framework created through agreed conventions in the measurement of distance and time intervals. And we must now bite the bullet and propose a definition of *mass*.* This Newton does, in the very first few words of Book I of *Mathematical Principles*:[4]

> *The quantity of matter is the measure of the same, arising from its density and bulk conjunctly....* It is this that I mean hereafter everywhere under the name body or mass. And the same is known by the weight of each body; for it is proportional to the weight, as I have found by experiments on pendulums, very accurately made, which shall be shewn hereafter.

We interpret Newton's use of the term 'bulk' to mean volume. So, the mass of an object is simply its density (the amount of mass per unit volume, measured—for example—in units of grams per

---

* So far I've only talked about weight. Now I'm switching to mass. Just to be clear, the weight of an object is what we measure using a scale of some kind, and depends on the mass of the object and the force of gravity. The weight of an object measured on the Moon will be much less than on Earth, because of the Moon's weaker gravity. However, the mass of the object is the same.

cubic centimetre) multiplied by the volume of the object (in cubic centimetres). This gives us a measure of the object's mass (in grams). This all seems perfectly logical and reasonable, although we'll soon be coming back to pick over this definition in an attempt to understand what it really says. Let's run with it for now.

So, an object has a certain intrinsic mass—the 'quantity of matter' in it—which is related to its density and volume, and which we can denote using the symbol m. Now, the property of mass that becomes its defining characteristic is that it is also the measure of the force or power of the object to *resist* changes in its state of motion. This is typically referred to as *inertial mass*. Why not simply call it 'mass' and acknowledge that all mass gives rise to inertia? We'll see why it becomes necessary to be a little more careful in our choice of terminology later in this chapter.

The evidence for the property of inertia comes from countless observations and measurements of objects both at rest (not moving) and in motion. I would argue that we cannot fail to gain at least qualitative evidence for the property of inertia just by living our daily lives. Newton drew on the body of accumulated experimental evidence (perhaps most notably the work of Galileo) to frame his first law: an object will persist in a state of rest (not moving) or of uniform motion in a straight line unless and until we apply (or 'impress') a force on it.[5]

Okay, so in addition to introducing and defining the concept of mass, Newton has now introduced a second concept: that of *force*. But what is a 'force', exactly? Newton explains that this is simply an 'action' exerted on an object that changes its state of motion, either from rest or from uniform motion in a straight line. The force *is* the action, and is impressed only for as long as the action lasts. Once the action is over, the force no longer acts on the object and it continues in its new state of motion until another force is impressed.[6]

There is no restriction on the kind of action (and hence the kind of force) that can be applied. I can kick the object; I can shoot it from a cannon; I can whirl it around like a discus; if it's made of conducting material I can charge it with electricity and move it with an electromagnet. All these different kinds of actions—all these different forces—act to change the state of motion of the object. This all seems clear enough, though there's a niggling doubt that these definitions are starting to look a bit circular. Let's keep going.

To get a real sense of what these definitions and the first law are telling us, we need to be a little clearer about what Newton means by 'motion', and specifically 'uniform motion in a straight line'. Let us imagine we are somehow able to set a perfectly spherical object in motion in a vacuum, with no external force acting on it. The first law tells us that this object will persist in its straight-line motion. If the sphere has an inertial mass m, and is set moving with a certain speed or velocity v (the rate of change in time of the object's position in space), then we determine the *linear momentum* of the object as its mass multiplied by velocity (m times v, or mv). This is what Newton means by 'uniform motion'.

Obviously, an object at rest has no velocity ($v = 0$) and therefore no linear momentum. The first law tells us that to get it moving, we need to impress a force on it. Likewise, when we impress a force on our spherical object with linear momentum mv, we change the momentum, by an amount determined by Newton's second law: the change in motion is proportional to the magnitude of the force and acts in the direction in which the force is impressed.[7]

What happens depends logically on the direction in which we apply the force. If we apply it in precisely the same direction in which the object is already moving, then we can expect to increase the linear momentum—we increase the motion, specifically the

speed with which the object is moving. If we apply it in the opposite direction, we will reduce the linear momentum. Applying a
force that exactly matches the linear momentum but in the
opposite direction will slow the object down and bring it back to
rest. Applying the force at an oblique angle may change the direction in which the object is moving.

We define the magnitude of the applied force F as the *rate* of
change of linear momentum that results from its impression.
The result is the famous statement of Newton's second law: $F = ma$,
force equals mass times acceleration.[8] Though famous, this result
actually does not make an appearance in the *Mathematical Principles*,
despite the fact that Newton must have seen this particular formulation of the second law in German mathematician Jakob
Hermann's treatise *Phoronomia*, published in 1716.* It is often
referred to as the 'Euler formulation', after the eighteenth-century
Swiss mathematician Leonhard Euler.

There is something deeply intuitively appealing about the
second law. Kick a stone with a certain force and it will fly
through the air, accelerated to some final speed before eventually succumbing to the force of gravity and returning to the
ground. A heavier stone with a greater inertial mass will require
a stronger kick to get the same acceleration. This description of
the subtle interplay between force, mass, and acceleration is
simple, yet profound. There can surely be no doubting its essential correctness.

Indeed, Newton's laws of motion have stood the test of time.
Yes, there are circumstances where we have to abandon them
because nature turns out to be even more subtle at the microscopic level of atoms and sub-atomic particles, and at speeds

---

* Newton issued a third edition of *Mathematical Principles* in 1726 and, if he had
been so minded, could have incorporated this version of the second law.

approaching that of light. We'll look at these circumstances in the chapters to follow. But for most of our 'everyday' applications on our earthly scale, Newton's laws work just fine.

I guess for the sake of completeness I should also state the third law: for every action we exert on an object (for every force we impress on it), there is an equal and opposite reaction.[9] Kick the stone, and the stone kicks back.

Now, if these laws are intended to provide a foundation for our understanding of the properties and behaviour of material substance, we should take a sober look at what they're telling us. The first thing we must do is acknowledge the inadequacy of Newton's definition of mass. As Mach, the arch-empiricist, explained, Newton defines mass as the quantity of matter in an object, measured by multiplying the density of the object by its volume. But we can only define the object's density in terms of its mass divided by its volume. This just takes us around in circles.[10] Defining mass in terms of density doesn't define mass at all.

But if we move on hurriedly and try to define mass using Newton's laws, we find we can't escape a vicious circularity. We could say that the inertial mass m is a measure of the resistance of an object to acceleration under an impressed force, F. But then what is the measure of the force F if not the extent to which it is resisted by the inertial mass m of the object? Try as we might, there's no way out.

The redeeming feature of $F = ma$ is that the impressed force F can result from many different kinds of actions. Whatever I choose to do, the measure of resistance to the different kinds of acceleration is the same. The inertial mass of the object appears to be consistently intrinsic to the object itself—it would seem to be a property of the object. But this doesn't take us any further forward. It doesn't tell us what mass *is*.

Mach sought to provide an *operational* definition of mass using the third law. This avoids the challenge by referring all measurements to a mass 'standard'. By working with ratios (e.g., of accelerations induced between two objects that are in collision), it is possible to eliminate force entirely and determine the ratio of the inertial masses of the two bodies. If one of these is an agreed standard, the inertial mass of the unknown body can then be measured in reference to it.

Let's pause to reflect on this. In the *Mathematical Principles*, Newton constructed the foundations for what we now call classical mechanics.* This is a structure that has proved itself time and again within its domain of applicability. Its central concepts of space and time, mass and force are intuitively appealing and appear to be consistent with our common experience. There is nothing in our familiar world of macroscopic objects moving in three-dimensional space that can't be explained using its laws supplemented, as appropriate, with similarly derived mechanical principles. Watch a game of tennis or snooker, and you'll quickly come to appreciate Newton's laws of motion.

Yet this is a structure that gives us nothing more than an operational definition of inertial mass, a definition that allows us to relate the properties of this object here with that object over there. It has nothing whatsoever to say about one of the most fundamental properties of all material objects. It doesn't tell us what mass *is*. And it gets worse.

Newton's ambition was not limited to the mechanics of earthly objects or, by inference, the mechanics of the atoms from which

---

* It is important to recognize these as 'foundations' and not to mistake them for the entire structure that was built on them. Solving practical, 'real-world' mechanical problems requires that Newton's three laws of motion be supplemented with many other principles, a task that absorbed the intellectual energies of many eighteenth-century mathematicians and physicists.

these objects might be composed. He sought to extend his science to describe the motions of heavenly bodies and, specifically, the planets. If successful, there would be no limits to the scope of such a science. It would be applied to matter in all its forms, from microscopic atoms to the familiar objects of everyday experience on Earth to objects in the furthest reaches of the visible universe. But to complete the picture he needed another law.

Now, unlike the laws of motion, in the *Mathematical Principles* Newton's famous inverse-square law of universal gravitation is not accorded the honour of a heading containing the word 'law'. The reasons for this are partly scientific, partly political. Newton communicated the draft of *Mathematical Principles* through the newly appointed Clerk to the Royal Society in London, the astronomer Edmond Halley. In May 1686, Halley wrote to Newton to tell him the good news: the Royal Society had agreed its publication.* But there was also bad news. At a recent Society meeting Robert Hooke—Boyle's erstwhile laboratory assistant, now Curator of Experiments at the Society—had claimed (loudly) that he, not Newton, had discovered the famous inverse-square law. Hooke had demanded that Newton acknowledge this.

The truth is that in 1681 Hooke had deduced the inverse-square relationship entirely empirically, from measurements he had performed with his assistant, Henry Hunt. But Hooke did not have the mathematical ability to construct a *theory* from which the inverse-square law could be derived.

It seems that Newton was inclined to be conciliatory, but as he received more news of Hooke's claims from other colleagues who had attended the meeting, his temper evaporated. He argued that mathematicians who 'do all the business' were being pressed

---

* As the Royal Society didn't have the funds to publish it, it seems Halley himself footed the bill.

into service as 'dry calculators and drudges'. Outraged by this attempt to undermine his own role in the discovery of the theory of universal gravitation, he dismissed Hooke as a man of 'strange, unsocial temper' and proceeded to reduce all acknowledgement to Hooke in the *Mathematical Principles* to the very barest minimum.[11]

So strong was his ire that he arranged the destruction of one (possibly two) portraits of Hooke that hung on the walls at the Royal Society's rooms. Consequently, there is no known portrait of Hooke, although the historian Lisa Jardine believed she had tracked down such a portrait in London's Natural History Museum, mislabelled as that of John Ray, a naturalist and Fellow of the Royal Society.

Newton was tempted to withdraw Book III (which he had called 'The System of the World') from the *Mathematical Principles* entirely. Instead, he replaced it with a section containing a very measured—and much more demanding—summary of observed astronomical phenomena followed by a sequence of propositions from which his 'system' can be deduced. Of the observed phenomena he described, Phaenomenon IV is the most worthy of note. At the time of writing, there were five known planets in addition to the Earth—Mercury, Venus, Mars, Jupiter, and Saturn. Newton noted that the time taken for each planet to complete its orbit around the Sun (the planet's orbital period) is proportional to the mean distance of the planet from the Sun. Specifically, the orbital period T is proportional to the mean distance r raised to the power $\frac{3}{2}$.[12]

After some unpacking, we can recognize this is as Kepler's third law of planetary motion. Based entirely on empirical observations of the motions of the planets that were known at the time measured relative to the fixed stars, Johannes Kepler had deduced that they move in elliptical orbits with the Sun at one focus, and that a line drawn from the Sun to each planet will sweep out equal areas in equal times as the planet moves in its orbit. The third law is a numerical relationship between the orbital period and the mean orbital distance or radius, such that T^2 is proportional to r^3 or, if we take the square root of both, T is proportional to $r^{3/2}$.[13]

Newton subsequently goes on to declare Proposition VII: there is a power or force of gravity intrinsic to all objects which is proportional to the quantity of matter that they contain.[14] It is then possible to use Kepler's third law to deduce the magnitude of the force of gravity of one object (such as a planet) acting on another, leading to Proposition VIII: in a system of two uniform spherical objects mutually attracted to each other by the force of gravity, the magnitude of the force varies inversely with the square of the distance between their centres.[15]

Most physics textbooks interpret these propositions as follows. If we denote the masses of two uniform spherical bodies 1 and 2 as m_1 and m_2, and r is the distance between their centres, then the force of gravity acting between them is proportional to $m_1 m_2 / r^2$. This is the famous inverse-square law. To complete this description, we need to introduce a constant of proportionality between this term and the measured force of gravity, which we again denote as F. This constant is a matter for measurement and convention. It is called the *gravitational constant* and is usually given the symbol G. It has a measured value of 6.674×10^{-11} Nm^2/kg^2 (Newton metres-squared per square kilogram).*

But this force of gravitation is just not like the forces generated by the various actions we considered in our discussion of the laws of motion. These latter forces are *impressed*; they are caused

---

* If the masses m_1 and m_2 are measured in kilograms and the distance r in metres, then the combination $m_1 m_2 / r^2$ has units kg^2/m^2 (square kilograms per square metre). Multiplying by G in units of Nm^2/kg^2 means that the gravitational force is calculated in units of N (Newtons).

by actions involving physical contact between the object at rest or moving in a state of uniform motion and whatever it is we are doing to change the object's motion. A stone will obey Newton's first law—it will persist in a state of rest—until I impress on it a force generated by swinging my right foot through the air and bringing it into contact with the stone. Or, to put it more simply, until I kick it.

But precisely what is it that is impressed upon the Moon as it swoons in Earth's gravitational embrace? How does the Moon (and the Sun) push the afternoon tide up against the shore? When a cocktail glass slips from a guest's fingers, what impresses on it and forces it to shatter on the wooden floor just a few feet below?

Newton was at a loss. His force of gravity seems to imply some kind of curious action-at-a-distance. Objects exert influences on each other over great distances through empty space, with nothing obviously transmitted between them. Critics accused him of introducing 'occult elements' into his theory of mechanics.

Newton was all-too-aware of this problem. In a general discussion (called a 'general scholium'), added in the 1713 second edition of *Mathematical Principles* at the end of Book III, he famously wrote: 'I have not been able to discover the cause of those properties of gravity from phenomena, and I frame no hypotheses.'[16] If that's not bad enough, we have no real evidence to support what might be an instinctive conclusion about the nature of the masses m_1 and m_2 which appear in the inverse-square law. We might be tempted to accord them the same status as the m in Newton's second law.

But physicists can be rather pedantic. In classical mechanics, motion and gravity appear to be different things, and some physicists prefer to distinguish gravitational mass (the mass responsible for the force of gravity) from inertial mass. Some go even further, distinguishing active gravitational mass (the mass responsible for *exerting* the force of gravity) from passive gravitational mass (the mass *acted on* by gravity).

For all practical purposes, empirical measures of these different kinds of mass give the same results, sometimes referred to as the 'Galilean' or 'weak' equivalence principle, and they are often used interchangeably. As we will see in Chapter 7, Einstein was willing to accept that these are all measures of an object's inertial mass. We must nevertheless accept an element of doubt.

Let's take stock once more. The research programmes of the mechanical philosophers had been building up to this. For sure, Newton didn't do all this alone. He 'stood on the shoulders of giants' to articulate a system that would lay foundations for an extraordinarily successful science of classical mechanics, one that would survive essentially unchallenged for 200 years. This was a science breathtaking in its scope, from the (speculative) microscopic atoms of material substance, to macroscopic objects of everyday experience, to the large-scale bodies of stars and planets.

However, look closely and we see that these foundations are really rather shaky. We have no explanation for the 'everyday' property of inertial mass, surely the most important property—we might say the 'primary' property—of material substance. We have no explanation for the phenomenon of gravity, another fundamentally important property of all matter. And without explanation, we must accept that we have no real *understanding* of these things.

We do have a set of concepts that work together, forming a structure that allows us to calculate things and predict things, enough to change profoundly the shape and nature of our human existence. This network of concepts is much like a game of pass the parcel. It's fun while the music plays, and we keep the parcel moving. But at some stage we know the music will stop.

Five things we learned

1. Newton defined the mass of an object in terms of its density and volume. But as density can only be defined in terms of mass and volume, this definition of mass is circular.
2. Newton also defined force in terms of an action that changes an object's state of motion—specifically, it changes the object's acceleration. Mass (or inertial mass) is then a measure of an object's resistance to changes in its acceleration. But we can't use this to define mass.
3. Newton's law of universal gravitation states that the force of gravity acting between two objects depends on their masses and the inverse-square of the distance between their centres.
4. But the force of gravity is not like the force that changes an object's state of motion. Gravity appears to work instantaneously, at a distance, with nothing obviously passing between the objects.
5. Despite the enormous success of Newton's system of mechanics, it does not really help us to understand what mass is.

4

THE SCEPTICAL
CHYMISTS

The different quantities of the same element contained in different molecules are all whole multiples of one and the same quantity, which, always being entire, has the right to be called an atom.

Stanislao Cannizzaro[1]

Newton was nothing if not ambitious. The *Mathematical Principles* set the foundations for a classical mechanics that could be applied to matter in all its forms. But missing from its scope was that other 'everyday' phenomenon, light. He set out to remedy this situation in 1704, with the publication of another treatise, titled *Opticks*.

At this time there were two competing theories of light. Newton's Dutch contemporary Christiaan Huygens had argued in favour of a wave theory in a treatise published in 1690. According to this theory, light is conceived to be a sequence of wave disturbances, with peaks and troughs moving up and down much like the ripples that spread out on the surface of a pond where a stone has been thrown.

But it's obvious that waves are disturbances *in* something. Throwing the stone causes a disturbance in the surface of the water, and it is waves *in the water* that ripple across the pond.

What, then, were light waves meant to be disturbances in? The advocates of the wave theory presumed that these must be waves in a tenuous form of matter called the *ether*, which was supposed to fill the entire universe.

Newton was having none of it. Tucked at the back of the first edition of *Opticks* is a series of sixteen 'queries', essentially rhetorical questions for which Newton provides ready answers. By the time of the publication of the fourth edition of *Opticks* in 1730, the number of queries had grown to thirty-one, the later additions each running to the length of a short essay. Much of *Opticks* is concerned with the properties of light 'rays', and in the very first definition Newton makes it clear that he regards these as discrete things: the 'least parts' of light.[2]

This leaves their status somewhat ambiguous, but in Querie 29 Newton makes clear what he thinks they really are: 'Are not the Rays of Light very small Bodies emitted from shining Substances?'[3] This 'atomic' description of light is rather less ambiguous, and has the virtue of bringing light within the scope of his mechanics.

But it is Newton's Querie 31 that allows us a further glimpse of his audacity. He observes that large objects act on each other through forces such as gravity, electricity, and magnetism, and he ponders on whether these or indeed other, unknown, forces of attraction might also be at work on the 'small particles of bodies'.[4]

Newton was ready to take the step that the Greek atomists had thought unnecessary. He was ready to replace the notion that the atoms interacted and joined together because of their shapes with the alternative idea that atoms interact and combine because of *forces* that act between them.* He refused to speculate on precisely

---

* In fairness to the Greeks, their atoms also 'swerved' and collided with each other, which I guess can be interpreted in Newton's mechanics in terms of forces acting between them.

what kinds of force might be involved, or how these might work. But in posing the question he opened the door to the idea that the motions of atoms and their combinations might be governed by forces that could be already known to us—gravity, electricity, and magnetism. Newton also dabbled in alchemy, and it is clear from the way that Querie 31 unfolds that he suspected that such 'atomic' forces, working with great strength at small distances, are responsible for the great variety of chemical reactions.[5]

We shouldn't get too carried away. Our hindsight shouldn't blind us to the simple truth that there was no empirical evidence to support Newton's suspicions. He had sought to account for already-known chemical facts based on an atomism now augmented by the notion of some form of inter-atomic force. This was an idea, not a theory, and he was consequently unable to use it to rationalize any existing chemical facts except in the most general terms. He was certainly unable to use the idea to make any testable *predictions*.

Any and all arguments relating to the properties and behaviour of atoms remained firmly limited to speculation, of the metaphysical kind. Although Newton's Querie 31 might seem to us tantalizingly prescient, there was nothing in it that could help the new breed of eighteenth-century scientists pursue these ideas any further through observation and experiment. They simply didn't have the wherewithal.

Instead, the founding fathers of modern chemistry simply went about their business, unravelling the mysteries of the deliciously complex chemical substances, their compounds, their combinations, and their reactions, that were now being revealed in the laboratory. They did this without much (if any) acknowledgement of the existence and likely role of the atoms of the mechanical philosophers. Their purpose was rather to try to make sense of this complexity and establish some order in it from the top down, and to develop and elaborate some fundamental (though entirely empirical) chemical principles based on what they could see, and do.

Central to the chemists' emerging logic was the concept of a *chemical element*. In the seventeenth and eighteenth centuries, the term 'element' had the same kind of position in the hierarchy of material substance as the ancient Greek elements earth, air, fire, and water. Of course, the chemists had by now determined that there was much more to material substance than these four elements. They had adapted the term to mean individual chemical substances that cannot be decomposed into simpler substances. And, although elements can be combined with other elements or compounds in chemical reactions, they do not thereby lose their identity, much as Boyle had explained in *The Sceptical Chymist*, first published in 1661.[6] The metallic silver, recovered in its 'pristine state' following the series of chemical manipulations described by Sennert, is an example of a chemical element.

Could chemical elements nevertheless be atoms? Recall that the atomic theory of Boyle and later of Newton afforded atoms spatial extension (although distinct shapes no longer featured), hardness, impenetrability, motion, and inertial mass. Querie 31 hinted at a possible role for some kind of force between them, but it didn't ascribe to them any *chemical* properties. It seems reasonable to suppose that, for those mechanical philosophers who dabbled in chemistry, it was anticipated that atoms sat somewhere even lower down in the hierarchy.

The unfolding history of eighteenth-century chemistry shows science doing what science does best: building a body of evidence that helps to dismantle previously received wisdom and replace it with new, more robust, ways of thinking about the world. However, this progress was (as it always is) rather tortuous. The chemists were practical men. (There were very few women engaged in this enterprise, except in supporting roles.) What they learned about chemistry would have very practical commercial implications and would help to establish the beginnings of the industrial revolution in the latter half of the eighteenth century. However, for the purposes of telling this story I will limit my choice of highlights to the efforts of the chemists to understand the different 'affinities' of the elements for each other, the 'rules' for combining them in chemical compounds, and thence to understand the nature of the elements themselves. As it happened, this turned out to be a lot easier when working with chemical substances in their gaseous form.

In the 1750s, whilst working as Professor of Anatomy and Chemistry at the University of Glasgow, Joseph Black discovered that treating limestone (calcium carbonate) with acids would liberate a gas which he called 'fixed air'. This is denser than regular air, and so would over time sink to the bottom of any vessel containing regular air. It would snuff out a candle flame, and the life of any animal immersed in it (don't ask). Passing the gas through a solution of limewater (calcium hydroxide) would cause calcium carbonate to be precipitated. We would eventually come to know this new gas as carbon dioxide.

The English chemist Joseph Priestley found that he could make fixed air by slowly dripping 'oil of vitriol' (sulphuric acid) onto a quantity of chalk (another form of calcium carbonate). In 1772, he published a short paper which explained how fixed air could be encouraged to dissolve in water, thereby producing 'carbonated water'.

He speculated on its possible medicinal properties, suggesting that it might be useful as a means to combat scurvy during long sea voyages. (He was wrong.) Nevertheless, like natural spring water, such artificially carbonated water has a pleasant taste. The German-born watchmaker and amateur scientist Johann Jacob

Schweppe used Priestley's methodology to develop an industrial-scale production process and founded the Schweppes Company in Geneva in 1783.

Priestley went on to perform a variety of observations and experiments on different 'kinds of air' (different gases), published in six volumes spanning the years 1774–1786. These included 'nitrous air' (nitric oxide), 'diminished nitrous air' (nitrous oxide), 'marine acid air' (hydrochloric acid), 'vitriolic acid air' (sulphur dioxide), and 'alkaline air' (ammonia). But it was his experiments on 'dephlogisticated air' that were to resonate in science history.

According to the theory of combustion that prevailed at the time, all combustible materials were thought to contain the element *phlogiston*, released when these materials burn in air. The theory had been established in 1667 by the German alchemist Johann Joachim Becher: the term phlogiston is derived from the Greek word for 'burning up'. So, when Priestley used a lens to focus sunlight on a sample of 'mercurius calcinatus per se' (mercuric oxide), he liberated a gas in which 'a candle burned…with a remarkably vigorous flame'.[7] Whatever this new gas was, it appeared to encourage a much more vigorous release of phlogiston, suggesting that it must be somehow more depleted of phlogiston than regular air. Priestley called it 'dephlogisticated air'.

Priestley's French contemporary, Antoine-Laurent de Lavoisier, disagreed with these conclusions. Lavoisier's approach differed somewhat from the more descriptive or qualitative approaches of his fellow chemists, bringing to his science elements of quantitative measurement and analysis more typical of the mechanical philosophers. Specifically, Lavoisier took pains carefully to *weigh* the substances he started with and the substances produced subsequently in a chemical reaction. For reactions involving gases, this meant using sealed glass vessels that could trap the gas and hold it secure. In some 1772 studies of the combustion of phosphorus and sulphur, he had observed that these substances *gain* weight as a result of burning. This was rather puzzling if burning involves the *release* of phlogiston to the air.

It seems that Lavoisier learned of 'dephlogisticated air' directly from Priestley himself, during the latter's visit to Paris in October 1774. He repeated and extended Priestley's experiments, and a few years later published a memoir in which he argued that combustion does not involve the release of phlogiston, but rather the chemical reaction of the combustible material with 'pure air', which is a component of regular air. He called it *oxygen*.

Combustion, then, is a process involving the *oxidation* of chemical substances, a process which may occur spontaneously or with the assistance of heat or light. In 1784, Lavoisier showed that oxygen could be reacted with 'inflammable air' (which he called hydrogen) to form water, thereby demonstrating unequivocally that a substance that had been regarded as an 'element' at least since Plato's *Timaeus* is, in fact, a chemical compound of hydrogen and oxygen.

In 1789, Lavoisier published *Traité Élémentaire de Chimie* ('Elementary Treatise of Chemistry'), widely regarded as the first textbook on 'modern' chemistry. It contains a list of chemical elements which includes hydrogen, nitrogen, oxygen, phosphorus, sulphur, zinc, and mercury, organized into 'metals' and 'non-metals'. The list also includes light and caloric (heat), still thought to be distinct elements at that time.

Alas, Lavoisier's story does not end happily. He was a powerful aristocrat, and an administrator of the *Ferme générale*, essentially a private customs and excise (or 'tax-farming') operation responsible for collecting taxes on behalf of the royal government. His work on combustion had earned him an appointment to the Gunpowder Commission, which came with a house and laboratory at the Royal Arsenal.

The French revolution changed the political order in 1789, and the rise of Maximilien Robespierre led to the Reign of Terror four years later. The *Ferme générale* was particularly unpopular with the revolutionaries, and an order for the arrest of all the former tax collectors was issued in 1793. Lavoisier was sent to the guillotine in May 1794. He was exonerated eighteen months after his execution.

Lavoisier's careful experiments had enabled him to establish an important principle. In a chemical transformation, mass (as measured by weight) is neither lost nor created: it is *conserved*. The total mass of the products of a chemical reaction will be the same as the total mass of the reactants. And, although they may become incorporated in different kinds of compounds as a result of some reaction, the identities of the chemical elements must also be conserved. This suggested rather strongly that the property of mass or weight might be traced to the individual chemical elements involved.

In October 1803, the English chemist John Dalton read a paper to the Manchester Literary and Philosophical Society, in which he famously observed that he had entered on the study of the relative weights of the 'ultimate particles of bodies', with remarkable success.[8] It seems that Dalton was inspired by aspects of Newton's atomism, but I'm going to take care to distinguish Dalton's 'ultimate particles', or *chemical atoms* (the atoms of chemical elements) from the physical atoms of Boyle and Newton. As I noted earlier, the chemical atoms have chemical properties that never formed part of the mechanical philosopher's atomic theory.

Dalton's work on relative weights culminated in the publication of his *New System of Chemical Philosophy* in 1808. This work features a table that now extends to twenty chemical elements, including carbon (or charcoal), sodium (soda), potassium (potash), copper, lead, silver, platinum (platina), and gold. The table also includes 'compound atoms' (which, to avoid confusion, I'm hereafter going to call *molecules*), consisting of combinations of between two and seven chemical atoms, assembled in whole-number ratios.

He devised an elaborate symbolism to represent the chemical atoms, for example using ⊙ for hydrogen and ○ for oxygen, and representing a molecule of water as the combination of one atom of hydrogen with one of oxygen, or ⊙○.

Dalton had focused on relative weight. His French contemporary Joseph Louis Gay-Lussac observed similar whole-number regularity in the *volumes* of gases that were combined. He found, for example, that two volumes of hydrogen would combine with one volume of oxygen to produce two volumes of water vapour. This just didn't fit Dalton's recipe. If we write 2⊙ + 1○ gives 2⊙○ we see that this chemical equation doesn't 'balance', there aren't enough oxygen atoms on the left-hand side. Dalton was dismissive.

The Italian scientist Amadeo Avogadro was sufficiently compelled by Gay-Lussac's work to turn these observations into a hypothesis (also sometimes referred to as Avogadro's law), which he published in 1811: equal volumes of all gases, at the same temperature and pressure, contain the same number of atoms or molecules.* But confusion still reigned. Avogadro noted that hydrogen and oxygen combine in the ratio 2:1 (not 1:1 as Dalton had assumed), but the combination produces *two* volumes of water vapour. This could only be possible if the oxygen atoms were somehow divisible. Few took any notice, and those who did weren't sure what to do with Avogadro's hypothesis.

A few years later, the Swedish physician and chemist Jöns Jacob Berzelius worked to sharpen and extend Dalton's system of atomic

---

* Confusingly, Avogadro referred only to molecules, calling what we would now call atoms 'elementary molecules'. Don't worry, we'll get there in the end.

weights and introduced a chemical notation which, with one modification, we still use today. Instead of Dalton's exotic symbols, Berzelius proposed to use a simple letter notation. Hydrogen is represented by H, oxygen by O. Berzelius wrote the 2:1 combination of these elements in water as H^2O. Today we write it as H_2O. Berzelius dodged the problem posed by the fact that two water molecules are produced by suggesting that Avogadro's hypothesis applies only to atoms, not molecules.

This is how science works. A detailed examination of science history shows that discoveries are very rarely—if ever—'clean', with a single individual scientist or a small group of collaborators leaping directly to the 'truth'. Instead, glimpses of the truth are often revealed as though through a near-impenetrable fog, with one scientist who has grasped part of the truth often arguing vehemently with another who has grasped a different part of the same truth. Progress is secured only when a sense of order can be established, and the fog is cleared.

Some historians have awarded Italian chemist Stanslao Cannizarro the role of clarifier, and there can be little doubting the clarifying nature of his treatise *Sketch of a Course of Chemical Philosophy*, published in 1858. At that time, Cannizzaro was Professor of Chemistry at the University of Genoa and had worked across all of chemistry's emerging sub-disciplines, physical, inorganic, and organic. He is known today largely for his 1851 discovery of the Cannizzaro reaction, involving the decomposition of a class of organic compounds called aldehydes to produce alcohols and carboxylic acids.

In the *Sketch*, Cannizzaro synthesized information available from studies of the densities of gases, specific heat capacities (the different capacities of substances to absorb and store heat), and the burgeoning chemistry of organic compounds, particularly in relation to the elucidation of their chemical formulae. All this information afforded opportunities to deduce an internally consistent set of atomic and molecular weights. But first, Cannizzaro had to establish that Avogadro's hypothesis makes sense only if we acknowledge the difference between atoms and molecules. He goes on:[9]

> 'Compare,' I say to them, 'the various quantities of the same element contained in the molecule of the free substance and in those of all its different compounds, and you will not be able to escape the following law: *The different quantities of the same element contained in different molecules are all whole multiples of one and the same quantity, which, always being entire, has the right to be called an atom.*'

The mystery of Gay-Lussac's measurements of the combining volumes of hydrogen and oxygen was now resolved. Setting the relative atomic weight of hydrogen equal to 1 based on hydrogen chloride (HCl), Cannizzaro observed that the relative weight of hydrogen present in hydrogen gas is actually 2. Hydrogen gas consists of molecules, not atoms: 'The atom of hydrogen is contained twice in the molecule of free hydrogen.'[10] Likewise, if the relative atomic weight of oxygen in water (H_2O) is taken to be 16, this 'quantity is half of that contained in the molecule of free oxygen'.[11]

Clearly, hydrogen and oxygen are both *diatomic gases*, which we write as H_2 and O_2, and the combining volumes now make perfect sense: $2H_2 + O_2 \rightarrow 2H_2O$. The equation balances—four hydrogen atoms in two molecules of hydrogen gas combine with two oxygen atoms in one molecule of oxygen gas to give two molecules of water.

Of course, all these elaborate combining rules don't *prove* the existence of chemical atoms, and some scientists remained stubbornly empiricist about them. But nevertheless when speculative theoretical entities are found to be useful across a range of scientific disciplines, over time scientists will inevitably start to invest belief in them.

In 1738, the Swiss physicist Daniel Bernoulli had argued that the properties of gases could be understood to derive from the rapid motions of the innumerable atoms or molecules in the gas. The pressure of the gas then results from the *impact* of these atoms or molecules on the surface of the vessel that contains them. Temperature simply results from the *motions* of the atoms or molecules. This *kinetic theory of gases* bounced around for a few decades before being refined by German physicist Rudolf Clausius in 1857. Two years later, Scottish physicist James Clerk Maxwell developed a mathematical formulation for the distribution of the velocities of the atoms or molecules in a gas. This was generalized in 1871 by Austrian physicist Ludwig Boltzmann, and is now known as the Maxwell–Boltzmann distribution.

The Maxwell–Boltzmann distribution can be manipulated to yield an estimate for the mean or most probable velocity of the atoms or molecules in a gas. For molecular oxygen (O_2) at room temperature, the mean velocity is about 400 metres per second. This is roughly the muzzle velocity of a bullet fired from an average rifle. (Fortunately for us, oxygen molecules are a lot lighter than bullets.)

This is all very well, but we still can't *see* these motions, and there may be other explanations for the properties of gases that have nothing to do with atoms or molecules. This final stumbling block was removed by Einstein in 1905, as I mentioned in Chapter 1. He suggested that tiny particles suspended in a liquid would be buffeted by the random motions of the molecules of the liquid. If the particles are small enough (but still visible through a microscope) it should be possible to see them being pushed around by the molecules (which remain invisible). He further speculated that this might be the explanation for the phenomenon of Brownian motion, but the experimental data were too imprecise to be sure.[12]

Better data were eventually forthcoming. French physicist Jean Perrin's detailed studies of Brownian motion in 1908 subsequently confirmed Einstein's explanation in terms of molecular motions. Perrin went on to determine that the value of what he suggested should be called Avogadro's constant, which scales the microscopic world of atoms and molecules to the macroscopic world in which we make our measurements.* There could now be no doubting the reality of atoms.

One small fly in the ointment: British physicist Joseph John Thompson had discovered and characterized a new particle, which he called the *electron* in 1895. New Zealander Ernest Rutherford discovered the *proton* in 1917. It seems that evidence for the existence of atoms was being established just as physicists were working out how to take these same atoms apart.

Five things we learned

1. Newton speculated that forces (of an unspecified nature) acting between the atoms of chemical substances are responsible for the great variety of chemical reactions.
2. Priestley and Lavoisier helped to clarify the nature of chemical substances in terms of the different chemical elements they contain, such as hydrogen, carbon, oxygen, etc.

---

* Avogadro's constant is 6.022×10^{23} 'particles' (such as atoms or molecules) per mole. The mole is a standard unit of chemical substance defined in relation to the number of atoms in 12 grams of carbon-12. Thus, a mole of oxygen (which has a volume of about 22 litres) will contain about 602 billion trillion molecules. That's a lot.

3. Dalton devised a system of chemical atoms, each with a different atomic weight. This was translated by Berzelius more or less into the system we use today, in which a molecule of water is represented as H_2O, where H and O represent hydrogen and oxygen atoms.

4. Cannizzaro clarified the rules for combining different atoms to form molecules. Hydrogen and oxygen are actually both diatomic gases, H_2 and O_2. In the reaction between hydrogen and oxygen, two molecules of hydrogen combine with one of oxygen to form two molecules of water: $2H_2 + O_2 \rightarrow 2H_2O$.

5. Einstein suggested, and Perrin demonstrated, that the Brownian motion of small particles suspended in a fluid is the result of the motions of the molecules of the fluid. It was finally accepted that atoms and molecules really do exist.

PART II

MASS AND ENERGY

In which time dilates, lengths contract, spacetime curves, and the universe expands. The mass of a body becomes a measure of its energy content, and most of the energy content of the universe is found to be missing.

5

A VERY INTERESTING CONCLUSION

The mass of a body is a measure of its energy content.

Albert Einstein[1]

As Chapter 4 implies, our understanding of the atomic and molecular nature of matter unfolded at speed in a relatively short time. For more than 2,000 years atoms had been the objects of metaphysical speculation, the preserve of philosophers. Over a period of fifty or sixty years beginning in the early nineteenth century, their status changed rather dramatically. By the early twentieth century, they had become the objects of serious *scientific* investigation.

The scientists sought to interpret the properties and behaviour of material substance using the structure of classical mechanics, erected on the foundations that had been laid down by Newton more than 200 years before. Small differences between theory and experiment were quite common, but could be reconciled by acknowledging that objects (including atoms and molecules) are more complex than the simplified models needed to make the theory easy to apply.

The simpler theories assume that atoms and molecules behave 'ideally', as though they are perfectly elastic point particles, meaning that they don't deform and don't occupy any volume in space.

They clearly do, and making allowances for this enabled scientists to take account of such 'non-ideal' behaviour entirely within the framework of classical mechanics.

But in the last decades of the nineteenth century this structure was beginning to creak under the strain of accumulating evidence to suggest that something was more fundamentally wrong. Many physicists (including Einstein) had by now developed deep reservations about Newton's conceptualization of an *absolute* space and time. These were reservations arguably born from a sense of philosophical unease (Mach, the arch-empiricist, rejected them completely), but they were greatly heightened by a growing *conflict* with James Clerk Maxwell's wave theory of electromagnetic radiation. Something had to give.

In the *Mathematical Principles*, Newton had been obliged to contemplate the very nature of space and time. Are these things aspects of an independent physical reality? Do they exist independently of objects and of our perceptions of them? As Kant might have put it, are they absolute things-in-themselves?

We might be tempted to ask: If space and time are not absolute, then what are they? The answer is: *relative*. Think about it. We measure distances on Earth relative to a co-ordinate system (e.g., of latitude and longitude). We measure time relative to a system based on the orbital motion of the Earth around the Sun and the spin motion of the Earth as it turns around its axis. These systems might seem 'natural' choices, but they are natural only for us Earth-bound human beings.

The simple fact is that our experience of space and time is entirely relative. We see objects moving towards or away from each other, changing their relative positions in space and in time. This is relative motion, occurring in a space and time that is in principle defined only by their relationships to the objects that exist within it. Newton was prepared to acknowledge this in what he called our 'vulgar' experience, but his system of mechanics demanded *absolute* motion. He argued that, although we can't directly perceive them, absolute space and time really must exist, forming a kind of 'container' within which actions impress forces on matter and things happen. Take all the matter out of the universe and the empty container would remain: there would still be 'something'. This was discomfiting to a few philosophically minded scientists perhaps, but it hardly seems something to be losing lots of sleep over.

Then along came Maxwell. We encountered the Scottish physicist James Clerk Maxwell towards the end of Chapter 4. Confronted by compelling experimental evidence for deep connections between the phenomena of electricity and magnetism, over a ten-year period from 1855 to 1865 he published a series of papers which set out a theory of *electrodynamics*. This theory describes electricity and magnetism in terms of two distinct, but intimately linked, electric and magnetic *fields*.

Now, there are many different kinds of examples of 'fields' in physics. Any physical quantity that has different magnitudes at different points in space and time can be represented in terms of a field. Of all the different possibilities (and we will meet quite a few in this book), the *magnetic field* is likely to be most familiar.

Think back to that science experiment you did in school. You sprinkle iron filings on a sheet of paper held above a bar magnet. The iron filings become magnetized and, because they are light, they shift position and organize themselves along the 'lines of force' of the magnetic field. The resulting pattern reflects the strength of the field and its direction, stretching from north to south poles. The field seems to exist in the 'empty' space around the outside of the bar of magnetic material.

The connection between electric and magnetic fields has some very important consequences. Pass electricity along a wire and you generate an electric current. You also create a changing magnetic field. Conversely, change a magnetic field and you generate an electric current. This is the basis for electricity generation in power stations. Maxwell's equations tie these fields together and explain how one produces the other.

There's more. When we look hard at Maxwell's equations we notice that they also happen to be equations that describe the motions of *waves*. Experimental evidence in support of a wave theory of light (which is one form of electromagnetic radiation) had been steadily accumulating in the years since Newton's *Optiks* was first published. When we squeeze light through a narrow aperture or slit in a metal plate, it spreads out (we say it *diffracts*), in much the same way that ocean waves passing through a narrow gap in a harbour wall will spread out in the sea beyond the wall. All that is required for this to happen is for the slit to be of a size similar to the average wavelength of the waves, the distance for one complete up-and-down, peak-to-trough motion.

Light also exhibits interference. Shine light on two slits side by side and it will diffract through both. The waves diffracted by each slit then run into each other. Where the peak of one wave meets the peak of the other, the result is constructive interference— the waves mutually reinforce to produce a bigger peak. Where trough meets trough the result is a deeper trough. But where peak meets trough the result is destructive interference: the waves cancel each other out. The result is a pattern of alternating brightness and darkness called *interference fringes*: bright bands are produced by constructive interference and dark bands by destructive interference. This is called two-slit interference (see Figure 2).

The waves 'bend' and change direction as they squeeze through the slits or around obstacles. This kind of behaviour is really hard to explain if light is presumed to be composed of 'atoms' obeying

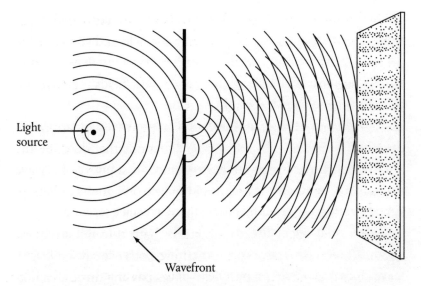

Light
source

Wavefront

Figure 2. When passed through two narrow, closely spaced apertures or slits, light produces a pattern of alternating light and dark fringes. These can be readily explained in terms of a wave theory of light in which overlapping waves interfere constructively (giving rise to a bright fringe) and destructively (dark fringe).

Newton's laws of motion and moving in straight lines. It is much easier to explain if we suppose light consists of waves.

In fact, Maxwell's equations can be manipulated to calculate the speed of electromagnetic waves travelling in a vacuum. It turns out that the result is precisely the speed of light, to which we give the special symbol c.* In 1856, the conclusion was inescapable. Light does not consist of atoms. It consists of electromagnetic waves.[2]

This all seems pretty conclusive, but once again we have to confront the tricky question: if light is indeed a wave disturbance, then what is the *medium* through which it moves? Maxwell didn't

---

* And, for the record, the speed of light in a vacuum is 299,792,458 metres per second or (2.998×10^8 metres per second).

doubt that light must move through the *ether*, thought to fill all of space. But if the ether is assumed to be stationary, then in principle it provides a frame of reference—precisely the kind of 'container'—against which absolute motion can be measured, after all. The ghost of Newton might have been unhappy that his theory of light had been abandoned, but he would surely have liked the ether.

Physicists turned their attentions to practicalities. If all space really is full of ether, then it should be possible to detect it. For sure, the ether was assumed to be quite *intangible*, and not something we could detect directly (otherwise we would already know about it). Nevertheless, even something pretty tenuous should leave clues that we might be able to detect, indirectly.

The Earth rotates around its axis at 465 metres per second. If a stationary ether really does exist, then the Earth must move through it. Let's just imagine for a moment that the ether is as thick as the air. Further imagine standing at the Earth's equator, facing west towards the direction of rotation. What would we experience? We would likely feel an *ether wind*, much like a strong gale blowing in from the sea which, if we spread our arms and legs, may lift us up and carry us backwards a few metres.* The difference is that the ether is assumed to be stationary: the wind is produced by Earth's motion through it.

Now, a sound wave carried in a high wind reaches us faster than a sound wave travelling in still air. The faster the medium is moving, the faster the wave it carries must move with it. This means that, although the ether is meant to be a lot more tenuous than the air, we would still expect that light waves carried along

---

* I'm being rather understated here. A hurricane force wind (with a Beaufort number of 12) has a speed anything above 32.6 metres per second. Let's face it, you really wouldn't want to be out in a wind blowing at 465 metres per second!

with the ether wind should travel faster than light waves moving against this direction. In other words, light travelling in a west-to-east direction, in which the ether wind is greatest, should be carried faster than light travelling east-to-west, in the opposite direction.

In 1887, American physicists Albert Michelson and Edward Morley set out to determine if such differences in the speed of light could be measured. They made use of subtle interference effects in a device called an interferometer, in which a beam of light is split and sent off along two different paths (see Figure 3). The beams along both paths set off 'in step', meaning that the position of the peaks along one path matches precisely the position of the peaks travelling along the other path. These beams are then brought back together and recombined. Now, if the total path taken by one beam is slightly longer than the total path taken by the other, then peak may no longer coincide with peak and the result is destructive interference. Alternatively, if the total paths are equal *but the speed of light is different* along different paths, then the result will again be interference.

No differences could be detected. Within the accuracy of the measurements, the speed of light was found to be constant. This is one of the most important 'negative' results in the entire history of science.

What's going on? Electromagnetic waves demand an ether to move in, yet no evidence for the ether could be found. These were rather desperate times, and in an attempt to salvage the ether physicists felt compelled to employ desperate measures. Irish physicist George FitzGerald (in 1889) and Dutch physicist Hendrik Lorentz (in 1892) independently suggested that the negative results of the Michelson–Morley experiments could be explained if the interferometer was assumed to be *physically contracting* along its length in response to pressure from the ether wind.

(a)

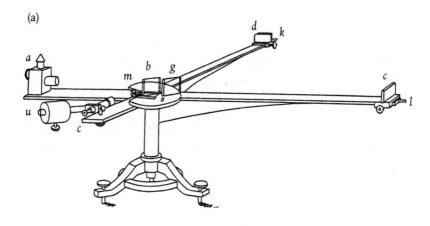

(b)

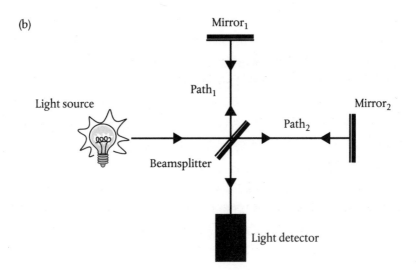

(c) At the beamsplitter

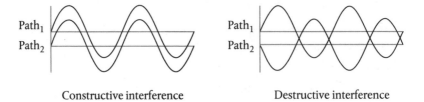

Constructive interference   Destructive interference

Figure 3. The Michelson–Morley experiment involved an apparatus called an interferometer, similar to the one shown here in (a). In this apparatus, a beam of light is passed through a half-silvered mirror or beamsplitter, as shown in (b). Some of the light follows path$_1$, bounces off mirror$_1$ and returns. The rest of the light follows path$_2$, bounces off mirror$_2$ and also returns. Light from both paths is then recombined in the beamsplitter and subsequently detected. If the light waves returning along both paths remain 'in step', with peaks and troughs aligned, then the result is a bright fringe (constructive interference). But if the lengths of path$_1$ and path$_2$ differ, or *the speed of light is different along different paths*, then the waves may no longer be in step, and the result is a dark fringe (destructive interference) as shown in (c).

Remember, interference should result if the light travelling along the path against the direction of the ether wind is carried a little slower. But if the length of this path is contracted by the 'pressure' of the ether wind, this could compensate for the change in the speed of light. The effects of a slower speed would be cancelled by the shortening of the distance, and no interference would be seen.

Fitzgerald and Lorentz figured out that if the 'proper' length of the path in the interferometer is l_0, it would have to contract to a length l given by l_0/γ. Here γ (Greek gamma) is the *Lorentz factor*, given by $1/\sqrt{(1 - v^2/c^2)}$, in which v is the speed of the interferometer relative to the stationary ether and c is the speed of light.

This factor will recur a few times in Chapters 6 and 7, so it's worthwhile taking a closer look at it here. The value of γ obviously depends on the relationship between v and c, as shown in Figure 4. If v is very much smaller than c, as, for example, in everyday situations such as driving to work, then v^2/c^2 is very small and the term in brackets is very close to 1. The square root of 1 is

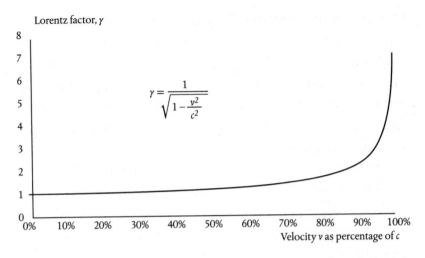

Figure 4. The Lorentz factor γ depends on the relationship between the speed at which an observed object is moving (v) and the speed of light (c).

1, so the Lorentz factor γ is also 1. At normal speeds your car doesn't contract due to the pressure of the ether wind.*

But now let's suppose that v is much larger, say 86.6 per cent of the speed of light (i.e., $v/c = 0.866$). The square of 0.866 (v^2/c^2) is about 0.75. Subtract this from 1 and we get 0.25. The square root of 0.25 is 0.50. So, in this case γ is equal to 2. A car travelling at this speed would be compressed to half its original length.

Now, the Earth's rotation speed is just 0.0002 per cent of the speed of light, so γ is only very slightly greater than 1. Any contraction in the path length in the interferometer was therefore expected to be very small.

There was no real explanation for such a contraction, and to some physicists it all looked like a rather grand conspiracy designed simply to preserve the idea of the ether and, by implication, absolute space. Einstein was having none of it. In the third

---

* Which is probably just as well.

of five papers that he published in 1905, he demolished the idea of the stationary ether and, by inference, absolute space.[3]

Einstein needed only to invoke two fundamental principles. The first, which became known as the *principle of relativity*, says that observers who find themselves in relative motion at different (but constant) speeds *must* make observations that obey precisely the same fundamental laws of physics. If I make a set of physical measurements in a laboratory on Earth and you make the same set of measurements aboard a supersonic aircraft or a spaceship, then we would expect to get the same results. This is, after all, what it means for a relationship between physical properties to be a 'law'.

The second principle relates to the speed of light. In Newton's mechanics, velocities are additive. Suppose you're on an inter-city train moving at 100 miles per hour. You run along the carriage in the same direction as the train at a speed of 10 miles per hour. We deduce that your total speed whilst running measured relative to the track or a stationary observer on a station platform is the sum of these, or 110 miles per hour.

But light doesn't obey this rule. Setting aside the possibility of a Fitzgerald–Lorentz-style contraction, the Michelson–Morley experiment shows that light always travels at the same speed. If I switch on a flashlight whilst on a stationary train the light moves away at the speed of light, c. If I switch on the same flashlight whilst on a train moving at 100 miles per hour, the speed of the light is still c, not c plus 100 miles per hour. Instead of trying to figure out *why* the speed of light is constant, irrespective of the motion of the source of the light, Einstein simply accepted this as an established fact and proceeded to work out the consequences.

To be fair, these principles are not really all that obvious. The speed of light is incredibly fast compared with the speeds of objects typical of our everyday observations of the world around us.

Normally this means that what we see appears simultaneously with what happens. This happens over here, and we see this 'instantaneously'. That happens over there shortly afterwards, and we have no difficulty in being able to order these events in time, this first, then that. Einstein was asking a very simple and straightforward question. However, it might appear to us, the speed of light is *not* infinite. If it actually takes some time for light to reach us from over here and over there, how does this affect our observations of things happening in space and in time?

Let's try to answer this question by performing a simple experiment in our heads.* Imagine we're travelling on a train together. It is night, and there is no light in the carriage. We fix a small flashlight to the floor of the carriage and a large mirror on the ceiling. The light flashes once, and the flash is reflected from the mirror and detected by a small light-sensitive cell or photodiode placed on the floor alongside the flashlight. Both flashlight and photodiode are connected to an electronic box of tricks that allows us to measure the time between the flash and its detection.

We make our first set of measurements whilst the train is stationary, and measure the time taken for the light to travel upwards from the flashlight, bounce off the mirror, and back down to the photodiode, as shown in Figure 5(a). Let's call this time t_0.

You now step off the train and repeat the measurement as the train moves past from left to right with velocity v, where v is a substantial fraction of the speed of light. Of course, trains can't move this fast in real life, but that's okay because this is only a thought experiment.

---

* Einstein was extremely fond of such 'thought experiments' (he called them *gedankenexperiments*). They're certainly a lot cheaper than real experiments.

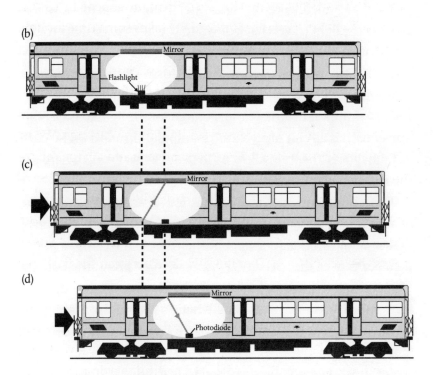

Figure 5. In this thought experiment, we measure the time taken for light to travel from the flashlight on the floor, bounce off the mirror on the ceiling, and return to the photodiode on the floor. We do this whilst the train is stationary (a), and record the time taken as t_o. You then observe the same sequence from the platform, but now as the train moves from left to right with a speed v, which is a substantial fraction of the speed of light (b)—(d). Because the train is moving, it now takes longer for the light to complete the round-trip from floor to ceiling and back, such that $t = \gamma t_o$, where γ is the Lorentz factor. From your perspective on the platform, time on the train appears dilated.

Now from your vantage point on the platform you see something rather different. The light no longer travels straight up and down. At a certain moment the light flashes, as shown in Figure 5(b). In the small (but finite) amount of time it takes for the light to travel upwards towards the ceiling, the train is moving forward, from left to right, Figure 5(c). It continues to move forward as the light travels back down to the floor to be detected by the photodiode, Figure 5(d).

From your perspective on the platform the light path now looks like a 'Λ', a Greek capital lambda or an upside-down 'V'. Let's assume that the total time required for the light to travel this longer path is t. A bit of algebraic manipulation and a knowledge of Pythagoras' theorem allow us to deduce that $t = \gamma t_0$, where γ is the Lorentz factor, as before.[4]

There is only one possible conclusion. From your perspective as a stationary observer standing on the platform, *time is measured to slow down on the moving train*. If the train is travelling at about 86.6 per cent of the speed of light, as we now know, the Lorentz factor $\gamma = 2$ and what took 1 second when the train was stationary now appears to take 2 seconds when measured from the platform. In different moving frames of reference, time appears to be 'dilated'.

I'm sure you won't be surprised to learn that distances measured in the direction of travel also contract, by precisely the amount that FitzGerald and Lorentz had demanded. But now the contraction is not some supposed physical compression due to the ether wind; it is simply a consequence of making measurements in moving frames of reference in which the speed of light is a universal constant.

You might be a little puzzled by this. Perhaps you're tempted to conclude: okay, I get that distances and times change when the object I'm making my observations on moves past me at different speeds, but isn't this a measurement problem? Surely there must be a 'correct' distance and a 'correct' time? Actually, no. There are only different frames of reference, including the so-called 'rest frame' of the object when it is stationary. If we're riding on the object looking out, any observations we make of other objects are subject to this relativity. There is no 'absolute' frame of reference, no 'God's-eye view'. *All* observations and measurements are relative.

It turns out that time intervals and distances are like two sides of the same coin. They are linked by the speed of the frame of reference in which measurements are made relative to the speed of light. It's possible to combine space and time together in such a way that time dilations are compensated by distance contractions, and vice versa. The result is a four-dimensional *spacetime*, sometimes called a *spacetime metric*. One such combination was identified by Hermann Minkowski, Einstein's former maths teacher at the Zurich Polytechnic. Minkowski believed that an independent space and time were now doomed to fade away, to be replaced by a unified spacetime.[5]

This is Einstein's *special theory of relativity*. At the time of its publication in 1905 the theory was breathtaking in its simplicity; the little bit of algebra in it isn't all that complicated, yet it is profound in its implications. But he wasn't quite done. He continued to think about the consequences of the theory and just a few months later he published a short addendum.

Before going on to consider what Einstein had to say next, we first need to update the story on Newton's second law concept of force. Whilst it is certainly true to say that this concept still has some relevance today, the attentions of nineteenth-century physicists switched from force to *energy*. This is the more fundamental concept. My foot connects with a stone, this action impressing a force upon the stone. But a better way of thinking about this is to see the action as transferring energy to the stone, in this case as an energy of motion (or what we call *kinetic energy*).

Like force, the concept of energy also has its roots in seventeenth-century philosophy. Leibniz wrote about *vis viva*, a 'living force' expressed as mass times velocity-squared—mv^2—which is only a factor of two larger than the expression for kinetic energy we use today ($\frac{1}{2}mv^2$). Leibniz also speculated that *vis viva* might be a *conserved* quantity, meaning that it can only be transferred between objects or transformed from one form to another—it can't be created or destroyed. The term 'energy' was introduced in the early nineteenth century and a law of conservation of energy was subsequently formulated, largely through the efforts of physicists concerned with the principles of thermodynamics.

In 1845, the ghost of 'caloric' was finally laid to rest when the English physicist James Joule identified the connection between heat and mechanical work. Heat is not an element, it is simply a measure of thermal (or heat) energy, and the amount of thermal energy in an object is characterized by its temperature. Spend ten minutes in the gym doing some mechanical work, such as lifting weights, and you'll make the connection between work, thermal energy, and temperature. Today the joule is the standard unit of energy, although the calorie—firmly embedded as a measure of the energy content of foodstuffs and an important dietary consideration—is the more widely known.*

Back to Einstein. He began his addendum with the words: 'The results of an electrodynamic investigation recently published by

---

* Actually, the 'calorie' of common usage is a 'large calorie', defined as the energy required to raise the temperature of 1 kilogram of water by 1°C. It is equal to about 4,200 joules.

me in this journal lead to a very interesting conclusion, which will be derived here.'[6] He wasn't kidding.

He considered the situation in which an object (e.g., an atom) emits two bursts of light in opposite directions, such that the linear momentum of the object is conserved. Each light burst is assumed to carry away an amount of energy equal to ½E, such that the total energy emitted by the object is E. Einstein then examined this process from two different perspectives or frames of reference. The first is the rest frame, the perspective of an observer 'riding' on the object and in which the object is judged to be stationary.

But we don't normally make measurements riding on objects. The second perspective is the more typical frame of reference involving a stationary observer (in a laboratory, say), making measurements as the object moves with velocity v, much as you stood on the platform making observations of the light beam on the moving train. Einstein deduced that the energy carried away by the bursts of light appears slightly larger in the moving frame of reference (it actually increases to γE), just as time appears dilated on the moving train.

But this process is subject to Einstein's principle of relativity. Irrespective of the frame of reference in which we're making our measurements, the law of conservation of energy *must* apply. So, if the energy carried away by the bursts of light is measured to be larger (γE) in the moving frame of reference compared with the rest frame (E), then that extra energy must come from somewhere.

From where? Well, the only difference between the two frames of reference is that one is stationary and the other is moving. We conclude that the extra energy can only come from the object's kinetic energy, its energy of motion. If we measure the energy carried away by the light bursts to be a little higher in the moving frame of reference, then we expect that the object's kinetic energy will be measured to be lower, such that the total energy is conserved.

This gives us two options. We know from the expression for kinetic energy—$\frac{1}{2}mv^2$—that the extra energy must come either from changes in the object's mass, m, or its velocity, v. Our instinct might be to leave the mass well alone. After all, mass is surely meant to be an intrinsic, 'primary' quality of the object. It would perhaps be more logical if the energy carried away by the bursts of light means that the object loses kinetic energy by *slowing down* instead.

Logical or not, in his paper Einstein showed that the velocity v is unchanged, so the object doesn't slow down as a result of emitting the bursts of light. Instead, the additional energy carried away by the light bursts in the moving frame of reference comes from the *mass* of the object, which falls by an amount $m = E/c^2$.[7] Einstein concluded:[8]

> If a body emits the energy [E] in the form of radiation, its mass decreases by [E/c^2]. Here it is obviously inessential that the energy taken from the body turns into radiant energy, so we are led to a more general conclusion: The mass of a body is a measure of its energy content.

Today we would probably rush to rearrange this equation to give the iconic formula $E = mc^2$.

Five things we learned

1. Newton's laws of motion require space and time to be considered absolute and independent of objects in the universe, in a kind of 'God's-eye view'.
2. Maxwell's equations describe electromagnetic radiation (including light) in terms of wave motion, but experiments

designed to detect the medium required to support such motion—the ether—came up empty.

3. These problems were resolved in Einstein's special theory of relativity, in which he was able to get rid of absolute space and time and eliminate the need for the ether.

4. We conclude that space and time are relative, not absolute. In different moving frames of reference time is measured to dilate and distances contract.

5. Einstein went on to use the special theory of relativity to demonstrate the equivalence of mass and energy, $m = E/c^2$. The mass of a body is a measure of its energy content.

6

INCOMMENSURABLE

Our analysis of the [Newtonian mass] vs. [relativistic mass] debate thus leads us to the conclusion that the conflict between these two formalisms is ultimately the disparity between two competing views of the development of physical science.

Max Jammer[1]

Needless to say, in the years since its publication, many physicists have picked over Einstein's derivation of this, his most famous equation. Some have criticized the derivation as circular. Others have criticized the critics. It seems that Einstein himself was not entirely satisfied with it and during his lifetime developed other derivations, some of which were variations on the same theme, others involving entirely different hypothetical physical situations.

Despite his efforts, all these different approaches seemed to involve situations that, it could be argued, are rather exceptional or contrived. As such, these different approaches are perhaps insufficiently general to warrant declaring $E = mc^2$ to be a deep, fundamental relationship—which Einstein called an 'equivalence'—between mass and energy.

This equation has by now become so familiar that we've likely stopped thinking about where it comes from or what it represents. So, let's pause to reflect on it here. Perhaps the first question we should ask ourselves concerns the *basis* of the equation. If this is supposed to be a fundamental equation describing the nature of material substance, why is the speed of light c involved in it? What has light (or electromagnetic radiation in general) got to do with matter? The second question we might ask concerns what the equation is actually telling us. It says that mass and energy are equivalent. But what does that *mean*, exactly?

Let's tackle these questions in turn. The basic form of $E = mc^2$ appears to tie the relationship between mass and energy to electrodynamics, the theory of the motion of electromagnetic bodies. Indeed, the first of Einstein's 1905 papers on special relativity was titled 'On the Electrodynamics of Moving Bodies', which is a bit of a giveaway.

For sure, any kind of physical measurement—hypothetical or real—will likely depend on light in some way. After all, we need to *see* to make our observations. But if this relationship is to represent something deeply fundamental about the nature of matter, then we must be able to make it more generally applicable. This means separating it from the motions of bodies that are electrically charged or magnetized and from situations involving the absorption or emission of electromagnetic radiation. Either we find a way to get rid of c entirely from the equation or we find another interpretation for it that has nothing to do with the speed of light.

This project was begun in 1909. American physicists Gilbert Lewis and Richard Tolman set out to establish a relativistic mechanics—meaning a mechanics that conforms to Einstein's special theory of relativity—that is general and universally applicable to all matter. They were only partially successful. They didn't quite manage to divorce c completely from its interpretation as the speed of light.

Their project was arguably completed only in 1972, by Basil Landau and Sam Sampanthar, mathematicians from Salford

University in England. Their derivation requires what appears to be a fairly innocent assumption; that *the mass of an object depends on its speed*. Landau and Sampanthar made no further assumptions about the precise nature of this dependence, and we can express it here simply as $m = f_v m_0$, where m is the mass of an object moving at velocity v, m_0 is the mass of the object when stationary (called the 'rest mass'), and f_v is some function of the velocity that has to be figured out.

What Landau and Sampanthar discovered as a consequence of their mathematical manipulations is that a quantity equivalent to c appears as a constant,* representing an absolute upper limit on the speed that any object can possess. The function f_v then becomes nothing other than the (by now hopefully familiar) Lorentz factor, γ.

What this suggests is that nothing—but nothing—in the universe can travel faster than this limiting speed. And for reasons that remain essentially mysterious, light (and, indeed, all particles thought to have no mass) travels at this ultimate speed. This answers our first question—we don't need to eliminate c from the equation connecting mass and energy; we just reinterpret it as a universal limiting speed.

But, as it so often turns out, solving one problem simply leads to another. The logic that Landau and Sampanthar applied would suggest that, just as space and time are relative, so too is the mass of an object. If $f_v = \gamma$, this means that $m = \gamma m_0$, where m is now the mass of an object in a frame of reference moving at velocity v and m_0 is the rest mass or 'proper mass', measured in the rest frame. An object moving at 86.6 per cent of the speed of light will appear to have a 'relativistic mass' m equal to twice the rest mass. It is measured to be twice as 'heavy'.

---

* For readers who know something of calculus, c appears as a constant of integration.

The object is not literally increasing in size. The m in $m = \gamma m_0$ represents mass as a measure of the object's inertia, which mushrooms towards infinity for objects travelling at or very near the limiting speed.* This is obviously impossible, and often interpreted as the reason why c represents an ultimate speed which cannot be exceeded. To accelerate any object with mass to the magnitude of c would require an infinite amount of energy.

But Einstein himself seems to have been rather cool on the idea of relativistic mass, and in certain physics circles the notion remains very dubious. In a 1948 letter to Lincoln Barnett, an editor at *Life* magazine who was working on a book about Einstein's relativistic universe, Einstein suggested that Barnett avoid any mention of relativistic mass and refer only to the rest mass.[2]

In an influential paper published in 1989, the Russian theorist Lev Okun reserved particular ire for the concept of relativistic mass. As far as he was concerned there is only one kind of mass in physics, the Newtonian mass m, which is independent of any frame of reference, whether moving or stationary.[3]

The simple truth is that even today there appears to be no real consensus among physicists on the status of these concepts. I have textbooks on special relativity sitting on my bookshelves which happily explore the consequences of the relativity of mass. I have other books and some papers stored on my computer which decry the notion and declare that there is only Newtonian mass, and special relativity is simply an extension of classical mechanics for the situation where objects move at speeds close to the ultimate limiting speed, c. The author of one textbook suggests that relativistic mass is a convenient construct and the

---

* Remember that as v gets very close to c, the ratio v^2/c^2 gets very close to 1 and γ increases rapidly towards infinity.

decision whether or not to use it is a matter of *taste*.[4] Let's park this for now and move on to our second question.

Today, nobody questions the fundamental nature of $E = mc^2$, or its essential correctness and generality. But, just as arguments have raged about the importance—or otherwise—of relativistic mass—so arguments have raged for more than 100 years about what the equivalence of mass and energy actually *means*. How come? Isn't it rather obvious what it means?

The default interpretation—one firmly embedded in the public consciousness, expressed in many textbooks and shared by many practicing physicists—is that $E = mc^2$ summarizes the extraordinary amount of energy that is somehow locked away like some vast reservoir *inside* material substance. It represents the amount of energy that can be liberated by the *conversion* of mass into energy.*

This was very much my own understanding, working as a young student in the 1970s, then as a postgraduate student and subsequently a university lecturer and researcher in chemical physics in the 1980s. To me, the fact that a small amount of mass could be converted into a large amount of energy just seemed to make matter—whatever it is—appear even more substantial.

In 1905, Einstein was quite doubtful that his 'very interesting conclusion' would have any practical applications, although he did note: 'It is not excluded that it will prove possible to test this theory using bodies whose energy content is variable to a high degree (e.g., radium salts).'[5]

The next few decades would provide many examples of bodies with highly variable energy content. Physicists discovered that the atoms of chemical elements have inner structures. Each atom

---

* If I'm not mistaken, my first encounter with the idea of 'mass into energy' was as a young child, watching a repeat of the 1953 BBC television series *The Quatermass Experiment*.

consists of a small central nucleus containing positively charged protons (discovered in 1917) and electrically neutral neutrons (discovered in 1932), surrounded or 'orbited' by negatively charged electrons (1895). It is the number of protons in the nucleus that determines the nature of the chemical element. Different elements, such as hydrogen, oxygen, sulphur, iron, uranium, and so on, all have different numbers of protons in their atomic nuclei. Atoms containing nuclei with the same numbers of protons but different numbers of neutrons are called *isotopes*. They are chemically identical, and differ only in their relative atomic weight and stability.

Physicists realized that the neutron could be fired into a positively charged nucleus without being resisted or diverted. Italian physicist Enrico Fermi and his research team in Rome began a systematic study of the effects of bombarding atomic nuclei with neutrons, starting with the lightest known elements and working their way through the entire periodic table. When in 1934 they fired neutrons at the heaviest known atomic nuclei, those of uranium, the Italian physicists presumed they had created even heavier elements that did not occur in nature, called *transuranic* elements. This discovery made headline news and was greeted as a great triumph for Italian science.

The discovery caught the attention of German chemist Otto Hahn at the prestigious Kaiser Wilhelm Institute for Chemistry in Berlin. He and his Austrian colleague Lise Meitner set about repeating Fermi's experiments and conducting their own, much more detailed, chemical investigations. Their collaboration was overtaken by events. When German forces marched into a welcoming Austria in the *Anschluss* of 12 March 1938, Meitner lost her Austrian citizenship. This had afforded her some protection from Nazi racial laws, but overnight she became a German Jew. The very next day she was denounced by a Nazi colleague and declared a danger to the Institute. She fled to Sweden.

Meitner celebrated Christmas 1938 with some Swedish friends in the small seaside village of Kungälv ('King's River') near Gothenburg. On Christmas Eve she was joined by her nephew, the physicist Otto Frisch. As they sat down to breakfast, Frisch had planned to tell his aunt all about a new experiment he was working on. However, he found that she was completely preoccupied. She was clutching a letter from Hahn, dated 19 December, which contained news of some new experimental results on uranium that were simply bizarre.

Hahn and another colleague Fritz Strassman had repeated Fermi's experiments and concluded that bombarding uranium with neutrons does not, after all, produce transuranic elements. It produces barium atoms. The most stable, common isotope of uranium contains 92 protons and 146 neutrons, giving a total of 238 'nucleons' altogether (written U-238). But the most common isotope of barium has just 56 protons and 82 neutrons, totalling 138. This was simply extraordinary, and unprecedented. *Bombarding uranium with neutrons had caused the uranium nucleus to split virtually in half.*

Meitner did a little energy book-keeping. She reckoned that the fragments created by splitting the uranium nucleus must carry away a sizeable amount of energy, which she estimated to be about 200 million electron volts (or mega electron volts, MeV).* The fragments would be propelled away from each other by the

---

* An electron volt is the amount of energy a single negatively charged electron gains when accelerated through a one-volt electric field. A 100W light bulb burns energy at the rate of about 600 billion billion electron volts per second. So, 200 million electron volts might not sound like much, but remember this is energy released by a *single nucleus*. A kilogram of uranium contains billions upon billions of nuclei. In fact, if every nucleus in a kilogram of uranium released 200 million electron volts of energy, this would be equivalent to the energy released by about 22 *thousand tons* of TNT.

mutual repulsion of their positive charges. Now, energy had to be conserved in this process, so where could it have come from?

She then recalled her first meeting with Einstein, in 1909. She had heard him lecture on his special theory of relativity, and had watched intently as he had derived his famous equation, $E = mc^2$. The very idea that mass could be *converted* to energy had left a deep impression on her. She also remembered that the nuclear masses of the fragments created by splitting a uranium nucleus would not quite add up to the mass of the original nucleus. These masses differed by about one-fifth of the mass of a single proton, mass that appeared to have gone 'missing' in the nuclear reaction. The sums checked out and it all seemed to fit together. A neutron causes the uranium nucleus to split almost in two, converting a tiny amount of mass into energy along the way. Frisch called it *nuclear fission*.*

The practical consequences of $E = mc^2$ would become all too painfully clear just a few years later. It turned out that it is the small quantity of the isotope U-235 present in naturally occurring uranium that is responsible for the observed fission, but the principles are the same. When scaled up to a 56-kilogramme bomb core of ninety per cent pure uranium-235, the amount of energy released by the disintegration of just a small amount of mass was sufficient to destroy utterly the Japanese city of Hiroshima, on 6 August 1945.†

---

* The history of the development and use of the world's first atomic weapons in recounted in my book *Atomic: The First War of Physics and the Secret History of the Atom Bomb* (Icon Books, London, 2009).

† I've deliberately chosen to work on this section of the book on 6 August, 2015. My daughter Emma is today in Hiroshima, where she will attend a screening of the film she has made whilst aboard a cruise organized by Peace Boat, a global, non-governmental organization based in Tokyo. The cruise carried a few *hibakusha*, survivors of the atomic bombings. The ship voyaged around the globe for 105 days in 2015, visiting twenty-three countries, spreading messages in support of a peaceful, nuclear-free world.

The destructive potential of $E = mc^2$ is clear. But, in fact, our very existence on Earth depends on this equation. In the so-called proton–proton (or p–p) chain which operates at the centres of stars, including the Sun, four protons are fused together in a sequence which produces the nucleus of a helium atom, consisting of two protons and two neutrons. If we carefully add up all the masses of the nucleons involved we discover a small discrepancy, called the *mass defect*. About 0.7 per cent of the mass of the four protons is converted into about 26 MeV of radiation energy, which when it comes from the Sun we call sunlight.*

This all seems pretty convincing. We import our classical preconceptions concerning mass largely unchanged into the structure of special relativity. We view the theory as an *extension* of classical mechanics to treat situations in which objects are moving very fast relative to the ultimate speed, which also happens to be the speed of light.

We discover that under exceptional circumstances, the mass of one or more protons can be encouraged to convert into energy. But this is still *Newtonian* mass. If we hang on to the concept of Newtonian mass (despite the fact that Newton himself couldn't really define it), then it seems we must reject the notion of relativistic mass. But if we reject relativistic mass, then Landau and Sampanthar's derivation is no longer relevant and we're stuck with the problem of how to interpret c. Einstein's most famous equation $E = mc^2$ seems so simple, yet we've managed to tie ourselves in knots trying to understand what it's telling us. Why?

We have to face up to the fact that we're dealing here with two distinctly different sets of concepts for which we happen to be

---

* Again, this might not sound too impressive. But this is the energy generated from the fusion of just four protons. It is estimated that in the Sun's core about 4×10^{38} protons react every second, releasing energy equivalent to about 4 million billion billion 100W light bulbs.

using the same terminology, and a degree of confusion is inevitable. In a Newtonian mechanics with an absolute space and time, mass is a *property* of material substance—perhaps an absolute or primary property, as the Greek atomists believed and as Locke described—and as such must indeed be independent of any specific frame of reference. In this conceptual framework we hold on to the dream of philosophers in its original form: material substance can be reduced to some kind of ultimate stuff, and the particles or atoms of this stuff possess the primary property of mass.

But in special relativity, in which space and time are relative, mass doesn't seem to behave like this. Mass is also relative. It does seem (or can at least be interpreted) to depend on the choice of reference frame and it appears to be intimately connected with the concept of energy, $m = E/c^2$.

So let's chart a different path, and see where it takes us. Einstein chose to title the 1905 addendum to his paper on special relativity thus: 'Does the Inertia of a Body Depend on its Energy Content?' This seems to me to be a very interesting choice of words.

As I explained in Chapter 5, Einstein offered a derivation which suggested that the energy E carried away by two bursts of light derives from the mass of the moving object, which decreases by an amount E/c^2. Now we can try to interpret this as something that happens to classical Newtonian mass under certain circumstances, involving a *conversion* of mass to energy, or we can instead assume that this is telling us *something completely new and fundamentally different about the nature of inertial mass itself.*

In this alternative interpretation, mass is *not* an intrinsic primary property of material substance; it is, rather, a *behaviour.* It is something that objects *do* rather than something that they *have.* Material substance contains energy, and it is this energy content which (somehow) gives rise to a resistance to acceleration, which we call inertia. We happen to have a tendency derived from a long association with classical mechanics (and, for that matter, everyday experience) to say that objects exhibiting inertia possess inertial mass.

In this conceptual framework we revise the dream of philosophers: material substance can be reduced to some form of energy, and this form of energy exhibits behaviour that we interpret as a resistance to acceleration. We could choose to eliminate mass entirely from this logic and from our equations and just deal with energy (though I'm confident that won't happen any time soon). The result would be 'mass without mass'—the behaviour we interpret as mass without the need for the property of mass. For sure, we still have some explaining to do. We would need to try to explain *how* energy gives rise to inertia.

But after working your way through the opening chapters of this book I suspect you're under no illusions. Trying to explain how energy creates inertia should be no more difficult in principle than doing the same for mass itself. After all, to say that mass is a measure of an object's inertia and that measurements of an object's inertia give us a handle on its mass is no explanation at all.

We still have conversion, but this is simply conversion of one form of energy into another. Of course, this just shifts the burden. Though it can take many forms, our intuition tells us that energy is a property: it is *possessed* by something. In this sense we tend to think about energy much like we think about temperature—not something that can exist independently of the things that possess it. The question now becomes: what is this something that possesses energy and exists in all material substance? Could this something still be called matter? We'll have the answers by the end of this book.

This is sometimes how science works. Very occasionally, we experience a revolution in scientific knowledge and understanding which completely changes the way we attempt to interpret theoretical concepts and entities in relation to empirical facts. Inevitably, the revolution is born within the old conceptual framework. Equations constructed in the old framework spring a few new surprises (such as $E = mc^2$), and the revolution begins. But through this process the old concepts tend to get dragged into the new framework, sometimes to serve the same purpose, sometimes to serve a different purpose.

The result is what Austrian philosopher Paul Feyerabend and American philosopher Thomas Kuhn called 'incommensurability'. Concepts dragged from the old framework are no longer interpreted in the same way in the new framework—strictly speaking they are no longer the same concept, even though the name hasn't changed. Some scientists hang on to the older, cherished interpretation. Others embrace the new interpretation. Arguments rage. The concepts are incommensurable.

And this, it seems, is what happened with mass. The old Newtonian concept of mass was dragged into the new framework of special relativity. Classical Newtonian mechanics is then seen as a limiting case of special relativity for situations involving speeds substantially less than that of c, the ultimate speed. But, for those willing to embrace the new interpretation, mass loses its primacy. It is just one (of many) forms of energy, a behaviour, not a property. Feyerabend wrote:[6]

> That the relativistic concept and the classical concept of mass are very different indeed becomes clear if we consider that the former is a *relation*, involving relative velocities, between an object and a co-ordinate system, whereas the latter is a *property* of the object itself and independent of its behavior in co-ordinate systems.

This would seem to make mass rather mysterious, or at least much more nebulous. But the simple truth is that—as we've seen—we never really got to grips with the classical conception of mass in the first place. Hand on heart, we never really understood it. Now we discover that it may not actually exist.

Five things we learned

1. In $E = mc^2$, the constant c is typically described as the speed of light, but it can be interpreted as an ultimate limiting speed which no object in the universe can exceed.

2. This interpretation of c as an ultimate speed relies on the assumption that mass is relative, just like space and time. But if this is true then this relativistic mass cannot be conceived in the same way as Newtonian mass.

3. $E = mc^2$ is typically interpreted to mean that mass can be *converted* to energy, as in an atomic bomb or in the proton–proton chain in the centres of many stars.

4. We can reconcile the notion of relativistic mass and the meaning of $E = mc^2$ by interpreting the mass of an object as a behaviour rather than a property; something the object *does* rather than something it *has*.

5. Dragging the concept of mass from Newtonian physics into special relativity has caused considerable confusion. The result is something that some philosophers of science have called incommensurability. Scientists argue at cross-purposes because they're interpreting mass in fundamentally different ways.

7

THE FABRIC

Spacetime tells matter how to move; matter tells spacetime
how to curve.

John Wheeler[1]

Einstein's special relativity was actually referred to as the
'theory of relativity' for a few years following its first appearance in 1905. It became 'special' when it was acknowledged that
the theory deals only with systems involving frames of reference moving at constant relative velocities. It doesn't cope
with frames of reference undergoing *acceleration*. And, because
Newton's force of universal gravitation is supposed to act *instantaneously* on gravitating bodies no matter how far apart they
might be, this classical conception of gravity is at odds with special relativity, which denies that the influence of any force can be
transmitted faster than the ultimate speed, *c* (and, by any measure, 'instantaneous' will always be faster). So, special relativity
can't describe objects undergoing acceleration nor can it describe
objects experiencing Newton's force of gravity.

Now, there are moments in the history of science when the interested onlooker can do nothing more than stare idiotically, jaw
dropped and mouth agape, at the sheer audacity that we have come
to associate with genius. Such a moment happened to Einstein in
November 1907, on an otherwise perfectly ordinary day at the

Patent Office in Bern. He had by this time received a promotion, to 'Technical Expert, Second Class'. As he later recounted: 'I was sitting in a chair in my patent office at Bern. Suddenly a thought struck me: If a man falls freely, he would not feel his weight.'[2]

Today we are so familiar with images and film clips of astronauts in zero-gravity environments that it may be difficult to grasp the immediate significance of Einstein's insight. But this seemingly innocent thought contains the seed of the solution that would unlock the entire mystery of Newton's force of gravity. Special relativity doesn't deal with acceleration or gravity, and Einstein now realized that these are not two problems to be solved, but one.

How come? Imagine you climb into an elevator at the top of the Empire State Building in New York City. You press the button to descend to the ground floor. Unknown to you, the elevator is, in fact, a disguised interstellar transport capsule built by an advanced alien civilization. Without knowing it, you are trans-ported instantaneously* into deep space, far from any planetary body or star. There is no gravity here. Now weightless, you begin to float helplessly above the floor of the elevator.

What goes through your mind? You have no way of knowing that you're now in deep space. As far as you're concerned, you're still in an elevator descending from the top of the Empire State Building. Your sensation of weightlessness suggests to you that the elevator hoist cables have been suddenly cut in some horrible accident. You're free-falling to the ground.

The alien intelligence observing your reactions does not want to alarm you unduly. They reach out with their minds, grasp the elevator/capsule in an invisible force field and gently accelerate it upwards. Inside the elevator, you fall to the floor. Relief washes over you. You conclude that the safety brakes must have engaged,

---

* Okay, so they're a *very* advanced civilization.

and you have ground to a halt. You know this because, as far as you can tell, you're once more experiencing the force of gravity.

Einstein called it the 'equivalence principle'. The local experiences of gravity and of acceleration are the same. They are one and the same thing. He called it his 'happiest thought'.[3]

But what does it mean? At first, Einstein wasn't entirely sure. It would take him another five years to figure out that the equivalence principle implies another extraordinary connection, between gravity and *geometry*. Despite how it might appear to us, the geometry of spacetime isn't 'flat' and rigid. It can bend and sag in places.

In school geometry classes, we learn that the angles of a triangle add up to 180°. We learn that the circumference of a circle is 2π times its radius, and that parallel lines never meet. Did you ever wonder why? These kinds of features (and many others besides) describe what mathematicians call a 'flat space', or 'Euclidean space', named for the famed geometer Euclid of Alexandria. This is the familiar three-dimensional space of everyday experience, on which we inscribe x, y, and z co-ordinate axes. When we combine this kind of space with the fourth dimension of time, such as Minkowski proposed, then the spacetime we get is a flat spacetime.

In a flat spacetime the shortest distance between two points is obviously the straight line we can draw between them. But what is the shortest distance between London, England, and Sydney, Australia? We can look up the answer: 10,553 miles. But this distance is not, in fact, a straight line. The surface of the Earth is curved, and the shortest distance between two points on such a surface is a curved path called an *arc of a great circle* or a *geodesic*. If you've ever tracked your progress on a long-haul flight, then this is the kind of path you would have been following.

Draw a triangle on the surface of the Earth (say by drawing lines between Reykjavik and Singapore, Singapore and San

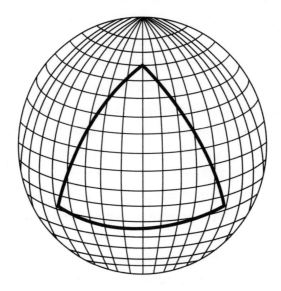

Figure 6. The angles of a triangle drawn on a sphere add up to more than 180°.

Francisco, San Francisco and Reykjavik) and you'll find its angles add up to more than 180° (see Figure 6). The circumference of a circle drawn on this surface is no longer equal to 2π times its radius. Lines of longitude are parallel at the equator but they meet at the poles.

Newton's first law of motion insists that an object will continue in its state of rest or uniform motion in a straight line unless acted on by an external force. In a flat spacetime all lines are straight, so Newton's force of gravity is obliged to act instantaneously, and at a distance. But, Einstein now realized, if spacetime is instead curved like an arc of a great circle, then an object moving along such a path will 'fall'. And as it falls, it accelerates.

In a curved spacetime it is no longer necessary to 'impress' the force of gravity on an object—the object quite happily slides down the curve and accelerates all on its own. All we need to do now is suppose that an object such as a star or a planet with a large mass-energy will curve the spacetime around it, much as a child curves the stretchy fabric of a trampoline as she bounces up and down. Other objects such as planets or moons that stray too close then follow the shortest path determined by this curvature. The acceleration associated with free-fall along this shortest path is then entirely equivalent to an acceleration due to the 'force' of gravity (see Figure 7).

American physicist John Wheeler summarized the situation rather succinctly some years later: 'Spacetime tells matter how to move; matter tells spacetime how to curve.'[4] Through this insight Einstein saw that he might now be able to account for both acceleration and gravity in what would become known as the *general theory of relativity*. What this theory suggests is that there is actually no such thing as the 'force' of gravity. Mass-energy does generate a gravitational field, but the field is space-time itself.

But, hold on. If gravity is the result of the curvature of space-time, and gravity is very much a part of our everyday experience

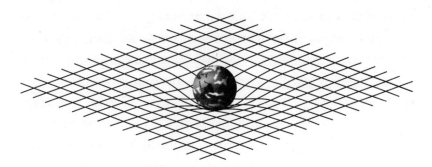

Figure 7. An object with a large mass-energy, such as the Earth, curves the spacetime around it. The effects of this curvature were studied by Gravity Probe B, a satellite mission which was launched in April 2004. The results were announced in May 2011, and provided a powerful vindication of the general theory of relativity.

here on Earth, then shouldn't we be able to *perceive* this curvature? Alas, the answer is no. The spacetime curvature caused by the mass-energy of the Earth is very slight, and subtle. The archetypal 'level playing field' on which we watch the game being played on a Sunday appears flat to us even though we know it sits on the surface of an Earth that is curved. In much the same way, our experience of spacetime is shaped by our local horizon. From our local perspective we perceive it to be flat even though we know it is gently curved. This is why we're still taught Euclidean geometry in school.

Now, Euclidean geometry is complicated enough, so when we add a fourth dimension of time we expect that things get more complicated still. It will come as no surprise to learn that the mathematics of curved spacetime involve an even higher level of abstraction.

We should note in passing that it is common to suppose that Einstein's genius extended to his ability in mathematics, and anyone flicking through a textbook on general relativity will marvel at the complexity arrayed in its pages. But Einstein was not particularly adept in maths. His maths teacher at the Zurich Polytechnic—Minkowski—declared him to be a 'lazy dog' and was greatly (but pleasantly) surprised by what he saw in Einstein's 1905 papers. Fortunately, as Einstein began to grapple with the abstract mathematics of curved spacetime, help was on hand in the form of a long-time friend and colleague, Marcel Grossman. '…you must help me, or else I'll go crazy', Einstein begged him.[5]

We can anticipate that Einstein's formulation of the general theory of relativity should be an equation (actually, a set of equations) of the kind which connect the curvature of spacetime—let's say on the left-hand side—with the distribution and flow of mass-energy on the right-hand side. Given a certain sizeable amount of mass-energy we need to be able to work out the extent to which the spacetime around it will curve and this, in turn, tells us how another quantity of mass-energy will be accelerated in response.

From his moment of inspiration in 1907 it took Einstein a further eight years to formulate his theory, with many false trails and dead ends. He eventually presented the field equations of general relativity to the Prussian Academy of Sciences in Berlin on 25 November 1915, just over 100 years ago. He later declared: 'The theory is beautiful beyond comparison. However, only one colleague has really been able to understand it....'[6] The colleague in question was German mathematician David Hilbert, who was in hot pursuit of the general theory of relativity independently of Einstein.

The resulting field equations were so complicated that Einstein judged them impossible to solve without making simplifying assumptions or approximations. And yet within a year the German mathematician Karl Schwarzschild had worked out a set of solutions. These are solutions for the specific case of a gravitational field outside a large, uncharged, non-rotating spherical body, which serves as a useful approximation for slowly rotating objects such as stars and planets.*

One of the more startling features of the Schwarzschild solutions is a fundamental boundary—called the *Schwarzschild radius*. Imagine a large spherical object, such as a star or planet. To escape the influence of the object's gravity, a rocket on the surface must be propelled at a speed which exceeds the object's *escape velocity*.[†] Further imagine that this object is compressed to a volume with a radius *smaller* than the Schwarzschild radius. To put this into some kind of perspective, note that the Schwarzschild radius of the Earth is about nine millimetres. Now the escape velocity is so

---

* Just to be clear—the Schwarzschild solutions are 'exact' for this kind of model or ideal situation, and it is the situation that serves as a useful approximation to real objects.

† The escape velocity of the Earth is a little over 11 kilometres per second.

large it *exceeds* the speed of light. Nothing—not even light itself—can escape the pull of the object's gravitational field. It's as though the spacetime is so distorted it has curved back on itself. The result is a *black hole*.*

Do black holes really exist in the universe? Although they are obviously difficult to detect directly, there is plenty of indirect evidence to suggest that black holes are fairly ubiquitous, and super-massive black holes are likely to sit at the centres of every galaxy.

From the beginning of his eight-year journey to the field equations of general relativity, Einstein was aware of four potential empirical tests. Actually, the first is not so much a test, more the resolution of a mystery. Newton's theory of universal gravitation predicts that planets should describe elliptical orbits around the Sun. In this description, the planet's point of closest approach to the Sun (called the *perihelion*) is a fixed point in space. The planet orbits the Sun and the perihelion is always at the same place.

However, observations of the orbits of the planets in the solar system show clearly that these points are not fixed. With each orbit the perihelion shifts slightly, or *precesses*. Anyone old enough to have played with a *Spirograph* drawing set as a child will appreciate the kinds of patterns that can result.

Of course, the Sun is not the only body in the solar system that generates a gravitational field. Much of the observed precession is caused by the cumulative gravitational pull of all the other planets. This contribution can be predicted using Newton's gravity. Collectively, it accounts for a precession in the perihelion of the planet Mercury of about 532 arc-seconds per century.[†]

---

* A name popularized (though not coined) by John Wheeler.

† A full circle is 360°, and an arc-minute is one-sixtieth of one degree. An arc-second is then one-sixtieth of an arc-minute. So, 532 arc-seconds represents about 0.15 of a degree.

However, the observed precession is rather more, about 574 arc-seconds per century, a difference of 42 arc-seconds.

Newton's gravity can't account for this difference and other explanations—such as the existence of another planet, closer to the Sun than Mercury (named Vulcan)—were suggested. But Vulcan could not be found. Einstein was delighted to discover that general relativity predicts a further contribution, of about 43 arc-seconds per century, due to the curvature of spacetime in the vicinity of Mercury.*

Perhaps the most famous prediction of general relativity concerns the bending of starlight passing close to the Sun. Now I'm not sure that everybody appreciates that Newton's gravity also predicts this phenomenon. After all, Newton originally conceived light in the form of tiny particles, each presumed to possess a small mass. Newtonian estimates suggest that light grazing the surface of the Sun should bend through 0.85 arc-seconds caused by the Sun's gravity.

The curvature of spacetime predicted by general relativity effectively doubles this, giving a total shift of 1.7 arc-seconds, and so providing a direct test. This prediction was famously borne out by a team led by British astrophysicist Arthur Eddington in May 1919. The team carried out observations of the light from a number of stars that grazed the Sun on its way to Earth.

Obviously, such starlight is usually obscured by the scattering of bright sunlight by the Earth's atmosphere. The light from distant stars passing close to the Sun can therefore only be observed during a total solar eclipse. Eddington's team recorded simultaneous observations in the cities of Sobral in Brazil and in São

---

* The perihelia of other planets are also susceptible to precession caused by the curvature of spacetime, but as these planets are further away from the Sun the contributions are much less pronounced.

Tomé and Príncipe on the west coast of Africa. The apparent positions of the stars were then compared with similar observations made in a clear night sky.

Although few could really understand the implications (and fewer still—even within the community of professional physicists—could follow the abstract mathematical arguments), the notion of curved spacetime captured the public's imagination and Einstein became an overnight sensation.

Incidentally, here's another answer to the question why we can't 'see' the curvature of spacetime. The mass of the Sun is about 330,000 times larger than the mass of the Earth, and yet the Sun curves spacetime only very slightly. I'd suggest a line that curves by only 1.7 arc-seconds—about 0.5 thousandths of a degree—would look exactly like a straight line to you or me.

General relativity also predicts effects arising from curved spacetime that are similar in some ways to the effects of special relativity. Einstein worked out the details in 1911, and they can be quickly identified using the Schwarzschild solutions.[7] Basically, to a stationary observer, time is measured to slow down (and distances are measured to contract) close to a gravitating object where the curvature of spacetime is strongest. A standard clock on Earth will run more slowly than a clock placed in orbit around the Earth.

Imagine a light wave emitted from a laboratory on Earth. The light wave is characterized by its wavelength—the distance covered by one complete up-and-down cycle. What happens to this wave as it moves away from the Earth and the effects of space-time curvature reduce?

Distances are shortened close to the Earth's surface where gravity is strong. Consequently, distances lengthen as we move away from Earth into outer space. A centimetre as measured on Earth's surface will be measured to be longer in space. The wavelength of a light wave will be measured to lengthen along with the distance. So, what happens is that as the light moves further and further away its wavelength is measured to get longer and longer.

A longer wavelength means that the light is measured to be 'redder' than it was. What we see as orange or yellow light on Earth, for example, will be measured to be red by the time it has reached a certain distance from Earth where gravity is much weaker. Physicists call this a 'redshift' or, more specifically, a *gravitational redshift*.

These are effects with some very practical consequences. A plane carrying an atomic clock from London to Washington DC loses 16 billionths of a second relative to a stationary clock left behind at the UK's National Physical Laboratory, due to time dilation associated with the speed of the aircraft. This is an effect of special relativity.

But the clock *gains* 53 billionths of a second due to the fact that gravity is weaker (spacetime is less curved) at a height of 10 kilometres above sea level. In this experiment, the net gain is therefore predicted to be about 40 billionths of a second. When these measurements were actually performed in 2005, the measured gain was reported to be 39±2 billionths of a second.[8]

Does this kind of accuracy really matter to anyone? The answer depends on the extent to which you rely on the Global Positioning System (GPS) used by one or more of your smartphone apps or your car's satellite navigation system. Without the corrections required by special and general relativity, you'd have a real hard time navigating to your destination.

Gravitational redshift happens as light moves away from a gravitating object. The opposite effect is possible. Light travelling towards a gravitating object will be blueshifted (shorter wavelengths) as the effects of gravity and spacetime curvature grow stronger.

The American physicists Robert Pound and Glen Rebka were the first to provide a practical Earth-bound test, at Harvard University in 1959. Electromagnetic waves emitted in the decay of radioactive iron atoms at the top of the Jefferson Physical Laboratory tower were found to be blueshifted by the time they reached the bottom, 22.5 metres below. The extent of the blueshift was found to be that predicted by general relativity, to within an accuracy of about ten per cent, reduced to one per cent in experiments conducted five years later.

There is one last prediction to consider. In June 1916, Einstein suggested that some kind of turbulent event involving a large gravitating object would produce 'ripples' in spacetime itself, much as we might feel the vibrations at the edge of a trampoline on which a child is bouncing. These ripples are called *gravitational waves*.

Despite how it might seem to us Earth-bound humans living daily with the force of gravity, this is actually nature's *weakest* force. Place a paper clip on the table in front of you. Gradually lower a handy-size magnet above it. At some point, the paper clip will be pulled upwards from the table by the magnetic force of attraction and will stick to the bottom of the magnet. You've just demonstrated to yourself that in the tussle for the paper clip, the magnetic force generated by the small magnet in your hand is stronger than the *gravitational pull of the entire Earth*.

This has important consequences. A spectacular event such as two black holes coalescing somewhere in the universe is predicted to produce gravitational waves which cause the spacetime nearby to fluctuate like a tsunami. But after travelling a long distance across the universe these waves are not only not noticeable (or we would have known about them long ago), they are barely detectable. Einstein was later quite dubious about them and it was not until the 1950s and 1960s that physicists thought they might stand a chance of actually detecting them.

The physicists have had to learn to be patient. But in September 2015 their patience was finally rewarded. About a billion years ago two black holes that had been orbiting each other rather warily in a so-called binary system finally gave up the ghost. They spiralled in towards each other and merged to form one supermassive black hole. The black holes had masses equivalent to twenty-nine times and thirty-six times the mass of the Sun, merging to form a new black hole with a mass sixty-two times that of the Sun. This event took place about 12 billion billion kilometres away in the direction of the Southern Celestial Hemisphere.

How do we know? We know because on 15 September 2015 the gravitational waves generated by this event were detected by an experimental collaboration called LIGO, which stands for Laser Interferometry Gravitational-wave Observatory. LIGO actually involves two observatories, one in Livingston, Louisiana and another at Hanford, near Richland in Washington, essentially on opposite sides of the continental United States.

Each observatory houses an L-shaped interferometer, with each arm measuring four kilometres in length. The principle is much the same as that described in Chapter 5, Figure 3 for the Michelson–Morley experiment, but on a much grander scale. The interferometers are set up in such a way that light interference causes a dark fringe at the detector, so that in a 'baseline' measurement no laser light is visible. Then, as the subtle 'movements' in spacetime caused by gravitational waves ripple through the interferometer, the lengths of the interferometer arms change: one is lengthened slightly and the other becomes shorter. The interference pattern shifts slightly towards a bright fringe, and the laser light reaching the detector becomes visible for a time. The changes in distances we're talking about here are breathtakingly small, measured in attometres (10^{-18} metres).

LIGO actually became operational in 2002 but was insufficiently sensitive. Operations were halted in 2010 and the observatories were upgraded with much more sensitive detectors. The enhanced observatories had only been operational for a couple of days when, quite by coincidence, the characteristic 'chirp' of gravitational waves from the black hole merger event passed through (Figure 8). This first successful detection of gravitational waves was announced at a press conference on 11 February 2016.

Of course, we were already pretty confident that general relativity is broadly correct, so the detection of gravitational waves wasn't needed to prove this. But this success is much, much more than a 'nice-to-know'. Now that we can detect gravitational waves we have a new window on events in distant parts of the universe, one that doesn't rely on light or other forms of electromagnetic radiation to tell us what's going on.

Let's reflect on all this before moving on. Einstein's gravitational field equations connect spacetime on the left-hand side with mass-energy and momentum on the right. They tell us how to calculate the shape of the gravitational field, or the 'gravitational well' that forms in spacetime around the Sun or the Earth, for example. From this shape we can determine how other objects in its vicinity will be obliged to move.

Figure 8. The LIGO observatory recorded its first gravitational-wave event on 14 September 2015. Signals detected at Hanford (H1) are shown on the left and Livingston (L1) on the right. Times are shown relative to 14 September 14 2015 at 09:50:45 UTC. Top row: the signals arriving first at L1 and then at H1 a few milliseconds later (for comparison, the H1 data are shown also on the right, shifted in time and inverted to account for the detectors' relative orientations). Second row: the signals as predicted by a model based on a black hole merger event. Third row: the difference between measured and predicted signals.

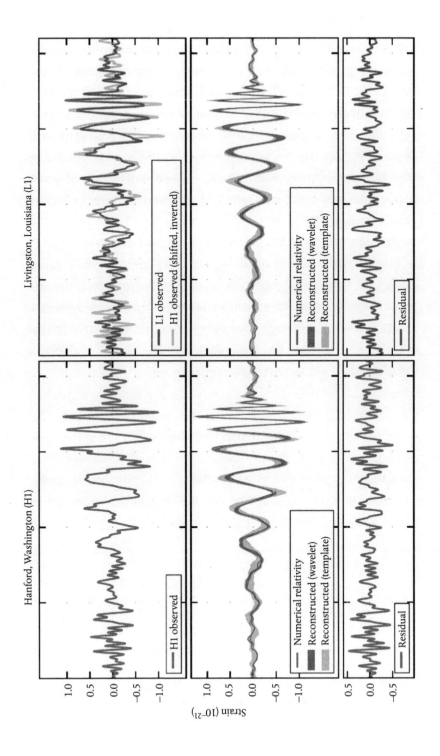

But, of course, the field equations do not tell us exactly *how* this is supposed to work. They provide us with a recipe for performing calculations, but they're rather vague on explanation. It might be best to think about it this way. Special relativity establishes two fundamental relationships between aspects of our physical reality that we had previously considered unconnected and independent of one another. Space and time meld into spacetime, and mass and energy into mass-energy. In general relativity, Einstein demonstrated that these too are in turn connected—spacetime and mass-energy are interdependent.

I like to think of it as the 'fabric' of our physical reality. This is a fabric of space, time, mass, and energy—all blurring somewhat around the edges—on which we attempt to weave a system of physics. Matter is energy, and only through approximation can it be separated from the space in which is sits, and the time in which it is.

Five things we learned

1. Special relativity is 'special' because it can't account for acceleration or Newton's force of gravity. In a flash of inspiration, Einstein realized that these are not two problems to be solved, but one.
2. The solution is to presume that in the vicinity of a large object such as a star or a planet, spacetime is curved, not flat.
3. Einstein's general theory of relativity connects the curvature of spacetime (on the left-hand side) with the density and flow of matter or mass-energy (on the right-hand side). Matter tells spacetime how to curve; spacetime tells matter how to move.
4. There have been many experimental tests of general relativity and many technologies that we have come to rely on depend

on its correctness. The detection of gravitational waves is a recent triumph for the theory.

5. Special relativity blends space and time into spacetime and mass and energy into mass-energy. General relativity connects mass-energy with the geometry of spacetime to form *the fabric*.

8

IN THE HEART OF DARKNESS

> We admittedly had to introduce an extension to the field equations that is not justified by our actual knowledge of gravitation.
>
> *Albert Einstein*[1]

A 'fabric' of space, time, mass, and energy is precisely what we need on which to weave not only a system of physics, but also an entire universe. So the obvious next question is: What do Einstein's field equations of general relativity tell us about the universe as a whole? Einstein himself furnished an answer in 1917, just two years after presenting his new theory to the Prussian Academy of Sciences.

At first sight, this seems like a case of theoretical physicists indulging in delusions of grandeur. How can a single set of equations be expected to describe something as complex as the whole universe? In truth, Einstein's field equations describe the relationship between spacetime and mass-energy, so in principle all the ingredients are present. It's just a question of *scale*.

Of course, the visible universe is full of lots of complicated physical, chemical, and biological things, such as galaxies containing gas, stars, black holes, quasars, neutron stars, planets, chemicals, and life. And, as we will soon discover, there is lots

These frequencies are fixed by the energy levels and the physics of the absorption or emission processes. They can be measured on Earth with great precision.

But if the light emitted by a hydrogen atom in a star that sits in a distant galaxy is moving relative to our viewpoint on Earth, then the spectral frequencies will be shifted by an amount that depends on the speed and direction in which the galaxy is moving.

Slipher discovered that light from the Andromeda nebula (soon to be re-named the Andromeda galaxy) is blueshifted (the light waves are compressed), suggesting that the galaxy is moving at a speed of about 300 kilometres per second *towards* the Milky Way. However, as he gathered more data on other galaxies, he found that most are redshifted (the light waves are stretched out), suggesting that they are all moving away, with speeds up to 1,100 kilometres per second.

In the 1920s, Hubble and Humason used the powerful 100-inch telescope at Mount Wilson near Pasadena, California, to gather data on more galaxies. What they found was that the majority of galaxies are indeed moving away from us. Hubble discovered what appears to be an almost absurdly simple relationship. The further away the galaxy, the faster it moves. This is *Hubble's law*.[5]

The fact that most of the galaxies are moving away from us does not place us in an especially privileged position at the centre of the universe. In an expanding universe it is spacetime that is expanding, with every point in spacetime moving further away from every other point. This means that the redshift in spectral frequencies is actually *not* caused by the Doppler Effect, after all, as this involves the motion of light *through* space. It is rather a *cosmological redshift*, caused by the expansion of the spacetime in which the light is travelling.

Think of the three-dimensional universe in terms of the two-dimensions of the skin of a balloon. If we cover the deflated balloon with evenly spaced dots, then as the balloon is inflated the dots all move away from each other. And the further away they are, the faster they appear to be moving.

Belgian physicist (and ordained priest) Abbé Georges Lemaître had actually predicted this kind of behaviour already in 1927, and had even derived a version of Hubble's law. In a subsequent paper published in 1933, Lemaître went on to suggest that the universe is expanding because empty spacetime is not, in fact, empty.[6] We can speculate that Aristotle would have been pleased.

Einstein had introduced his cosmological term on the left-hand side of his field equations, as a modification of spacetime itself, designed to offset the effects of the curvature caused by all the mass-energy on the right. But it takes just a moment to move the term across to the right-hand side of the equation. Now it represents a *positive* contribution to the total mass-energy of the universe.

This is not the familiar mass-energy we associate with stars and planets. It still depends on the structure of spacetime and suggests that 'empty' spacetime has an energy, sometimes called *vacuum energy*. In fact, Λ is directly proportional to the *density* of the vacuum energy, the amount of energy per unit volume of 'empty' spacetime.

If the universe is expanding, then simple extrapolation backwards in time would suggest that there must have been a moment in its history when all the energy in the universe was compacted to an infinitesimally small point, from which it burst forth in what we now call the 'big bang'. For a time it was thought that the subsequent evolution of the universe could be adequately described in terms of the tug-of-war between the post-big bang expansion rate and the mass-energy contained in the universe. Einstein's 'fudge' and the vacuum energy it implied, was deemed to be unnecessary and was quietly forgotten.

Now, we don't have a valid scientific theory that describes the 'beginning' or the very earliest moments in the evolution of the universe. But by making some assumptions about its size and temperature as it expands, we can with some confidence apply currently accepted theories to understand what might have happened to the contents of the universe, right back to about a trillionth of a second after the big bang. I think this is pretty impressive.

The constituents of the universe quickly combine to produce protons and neutrons, within about a second or so after the big bang. After about 100 seconds, protons and neutrons combine to form mostly helium nuclei (consisting of two protons and two neutrons, accounting for twenty-four per cent of the total by mass) and a thin scattering of heavier elements. Unreacted hydrogen nuclei (free protons) account for seventy-six per cent by mass.

We have to wait about 380,000 years for the further expansion of the universe to cause the temperature to fall to about three thousand degrees, at which point electrons combine first with helium nuclei, then hydrogen nuclei, to form neutral atoms in a process called 'recombination'. Up until this point the universe had been quite opaque, with light bouncing back and forth between the electrically charged nuclei and electrons, in a plasma that produced an impenetrable 'fog'.

But when the first neutral hydrogen and helium atoms are formed through recombination, there is nowhere else for the light to go and a flood of hot electromagnetic radiation is released into the universe. This radiation is cooled as the universe continues to expand, and appears today in the form of microwaves and infrared radiation with an average temperature of around −270.5°C, or 2.7 kelvin,* almost three degrees above absolute

---

* In the Celsius (or centigrade) temperature scale water freezes at 0°C and boils at 100°C. In the kelvin temperature scale zero kelvin corresponds to absolute zero (the ultimate zero on the temperature scale), or about −273.15°C.

zero. It is a cold remnant, the 'afterglow' of an eventful moment in the history of the universe. It is called the *cosmic background radiation*.

This radiation was first detected in 1964, by radio astronomers Arno Penzias and Robert Wilson, working at the Bell Laboratories' Holmdel research facility in New Jersey. A succession of satellite surveys—COBE (launched in 1989), WMAP (2001), and Planck (2009)—has mapped its subtle temperature variations in ever more exquisite detail.* Analysis of these maps provides much of

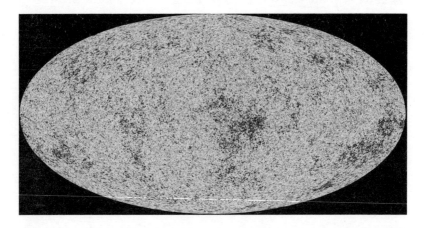

Figure 9. The detailed, all-sky map of temperature variations in the cosmic background radiation derived from data obtained from the Planck satellite. The temperature variations are of the order of ±200 millionths of a degree and are shown as false-colour differences, with lighter colours indicating higher temperatures and darker colours indicating cooler temperatures.

---

* COBE stands for COsmic Background Explorer. WMAP stands for Wilkinson Microwave Anisotropy Probe, named for David Wilkinson, a member of the COBE team and leader of the design team for WMAP, who died in 2002 after a long battle with cancer. The Planck satellite is named for German physicist Max Planck.

the observational evidence on which theories of the origin and evolution of the universe are constructed (see Figure 9).

You might be tempted to think that we now have all the ingredients required to build a universe, in a hot big bang. But you couldn't be more wrong. The shapes of spiral galaxies like the Andromeda galaxy (and, by inference, our own Milky Way) are not only very beautiful, they are also quite suggestive. We have no real difficulty accepting that such galaxies are *rotating*, trailing a swirl of stars, dust, and gas from their spiral arms. They are indeed rotating, and by observing stars at different distances from the galactic centre we can calculate the speeds with which these stars are orbiting about the centre.

What might we expect to see? Well, our own solar system provides a practical model in which most of the mass is concentrated at the centre (in the Sun) around which orbit the planets. Planets close to the centre feel the full force of the Sun's gravity and are whirled around at high speeds. Mercury, with an average orbital radius of about 58 million kilometres, has an orbital speed of nearly 50 kilometres per second. But as we get further and further away from the centre, the effects of the Sun's gravitational field weaken, and so we would expect distant planets to 'drag their feet' and orbit more slowly. Neptune, with an average orbital radius of about 4,500 million kilometres, has an orbital speed of only 5.4 kilometres per second.

A spiral galaxy is a little more complicated. The population of stars is indeed densest at the centre of the galaxy and if we accept the suggestion that super-massive black holes are also be to found here, then we can conclude that the force of gravity (the curvature of spacetime) is greatest here, too. But there are also lots and lots of stars orbiting at considerable distances from the centre, each adding to the gravitational field. Instead of peaking at a fairly sharp point right at the centre and falling away, the field is more like a Bell curve: it falls away gently at first, more strongly as the distance increases. The upshot of this is that we expect the *rotation curve*—the graph of orbital speed versus radius—to rise to a peak close to the centre before falling away. But when we measure the orbital speeds of stars at various distances from the centre, we find the rotation curve looks completely different. Instead of the orbital speeds declining with distance, they either flatten off or actually *increase* gently with increasing distance.

This effect was first observed in 1934 by the Swiss astronomer Fritz Zwicky, for a cluster of galaxies rather than a single galaxy. The higher orbital speeds at large distances must mean that gravity is much stronger here than we would anticipate from the mass of the stars that we can see. He concluded that as much as ninety per cent of the mass required to explain the orbital speeds appeared to be 'missing', or invisible. He called it 'missing mass'.

In 1975, American astronomer Vera Rubin and her colleague Kent Ford measured the speeds of stars orbiting single galaxies. They observed precisely the same effect—orbital speeds at the edge of the galaxy are much higher than predicted based on the mass of its visible stars. The rotation curves can only be explained by presuming that each galaxy sits at the centre of a 'halo' of missing or invisible matter, which now goes by the name 'dark matter'.

Dark matter is mysterious, but it is nevertheless profoundly important. The concentration of visible matter at the centres of giant dark matter halos is believed to have been absolutely essential in enabling the first stars and galaxies to form, a few hundred million years after the big bang. But we have absolutely no idea what it is. And we're still not quite done.

In the early 1990s, cosmologists wrestled with another stubborn problem. As we have already discussed in Chapter 7, despite the curvature caused by the Earth's gravity our local experience is that spacetime is flat, or Euclidean. But, aside from curvature caused by things like black holes, galaxies, individual stars, and planets, our observations are consistent with a universe in which spacetime as a whole is also broadly flat. Cosmologists understood that a flat spacetime could only be achieved if there is a very fine balance between the expansion of spacetime and the amount of mass-energy in it.

A so-called 'closed' universe contains a high density of mass-energy which will slow the expansion and eventually reverse it, causing the universe to collapse back in on itself. The spacetime of such a universe would be expected to be positively curved, like the surface of a sphere, on which the angles of a triangle add up to more than 180°.

An 'open' universe contains insufficient mass-energy to prevent the universe from expanding forever, and hence a spacetime which is expected to be negatively curved, like the surface of a saddle. It seems we live in a universe which exhibits a very fine balance between expansion rate and mass-energy, and so has a flat spacetime.

How much mass-energy do we need? We can use currently accepted cosmological parameters to estimate the value of the critical average density required for our kind of universe. When we plug the numbers in, we find that this critical density is about 8.6×10^{-30} grams per cubic centimetre.[7]

Let's put this into some kind of perspective. This density corresponds to an average of roughly 7 million protons in a volume of space the size of St Paul's Cathedral in London.[8] This might sound like a lot, but if we count up the protons and neutrons in the volume of air that normally fills the Cathedral, we get about 9.6 hundred billion billion trillion (9.6×10^{32}). Clearly, matter is a lot more concentrated in some places in the universe than others.

The problem was this. Cosmologists simply couldn't find enough mass-energy in the universe to ensure a flat spacetime. Even dark matter didn't help. In the early 1990s, the observed (and implied) mass of the universe was only of the order of thirty per cent of the total needed (the equivalent of only 2 million protons in St Paul's Cathedral). If this really is all there is, then the universe should be 'open', and spacetime should be negatively curved. What was going on?

There weren't that many options. Some astrophysicists realized that Einstein's fudge factor might after all have a role to play in the equations of the universe. A mass-energy density contributing thirty per cent and a cosmological constant contributing a vacuum energy-density of seventy per cent might be the only way to explain how the universe is flat. The vacuum energy became more familiarly known as 'dark energy'.

Different densities of mass-energy (of visible and dark matter and dark energy) predict different expansion rates at different times in the history of the universe. So, if we could determine the *actual* expansion history, then we'd be able to get a fix on what the actual density is. At first sight, this might not seem to help much. But because the speed of light is finite, when we look at events in very distant parts of the universe we see these as they would have appeared at the time the light was emitted. Looking at distant events allows us to look back into the history of the universe.

For example, light from the Sun shows us how the Sun looked about eight minutes ago, which is the amount of time it takes for the light to travel 150 million kilometres to the Earth. Light from the Andromeda galaxy shows us how it was 2.5 million years ago. Looking at galaxies that emitted their light many billions of years ago provides us with potential clues about how the universe was expanding then.

But the most distant galaxies are also the faintest, making it very difficult to measure their distances. This is indeed a problem, until one of its stars explodes spectacularly in a supernova.

Then, for a brief time, the galaxy lights up like a beacon in the darkness.

In 1998, two independent groups of astronomers reported the results of measurements of the redshifts of very distant galaxies illuminated by a certain type of supernova. These were the Supernova Cosmology Project (SCP), based at the Lawrence Berkeley National Laboratory near San Francisco, California, headed by American astrophysicist Saul Perlmutter, and the High-z (high-redshift) Supernova Search Team formed by US-born Australian-American Brian Schmidt and American Nicholas Suntzeff at the Cerro Tololo Inter-American Observatory in Chile. The results reported by both groups suggest that, contrary to the expectations that prevailed at the time, we live in a universe in which the expansion of spacetime is actually *accelerating*.

Dark energy was the only explanation. And, as dark energy is equivalent to a cosmological constant, Einstein's factor was officially no longer a fudge.

Consensus has now gathered around a version of big bang cosmology called variously the 'concordance' model, the 'standard model of big bang cosmology', or the Λ-CDM model, where Λ stands for the cosmological constant and CDM stands for 'cold, dark matter'. The model is based on a spacetime metric which provides an exact solution to Einstein's field equations and describes a homogeneous, expanding universe. It is known as the Friedmann–Lemaître–Robertson–Walker (FLRW) metric, named for Russian mathematician Alexander Friedmann, Georges Lemaître, American physicist Howard Robertson, and English theorist Arthur Walker.

The Λ-CDM model has six parameters which are adjusted to ensure consistency with observations (e.g., of the temperature variations in the cosmic background radiation, and the supernova redshift measurements, among others). Three of these parameters are the age of the universe, the density of dark matter, and the density of visible matter, the stuff of gas clouds, stars, planets, and us. Other parameters, such as the density of dark energy, can be deduced from these.

These parameters can be determined with some precision. The best-fit results from analysis of the Planck satellite data published in February 2015 suggests that the universe began 13.8 billion years ago. Dark energy accounts for about 69.1 per cent of the energy density of the universe and dark matter 26.0 per cent. The visible matter, or what we used to think of as 'the universe' not so very long ago, accounts for just 4.9 per cent. This means that the evolution of the universe to date has been determined largely by the push-and-pull between the anti-gravity of dark energy and the gravity of (mostly) dark matter.[9] Visible matter is carried along for the ride.

We've reached a very interesting moment in the history of our scientific understanding of the universe. We have a model based on Einstein's general theory of relativity which is simply extraordinary in its ability to accommodate all the observations of modern cosmology, astronomy, and astrophysics. And yet, at the same time, the model reveals an extraordinary level of ignorance. Although we can measure its gravitational effects, and we know that the universe we live in cannot have happened without it, we do not know what dark matter is. Although we observe the rate of expansion of the universe to be accelerating, we can't account for the existence of dark energy and we don't really know what this is, either. In other words, we have no real explanation for more than ninety-five per cent of the known universe.

Five things we learned

1. By making some simplifying assumptions, Einstein was able to apply his general theory of relativity to the entire universe.

He found he needed to add an additional term to his equations which includes a 'cosmological constant'.

2. Observations of distant galaxies showed that most galaxies are moving away from us with speeds that are proportional to their distances (Hubble's Law). The universe is expanding, and extrapolation backwards in time suggests that it began in a 'big bang'.

3. The rotation curves of galaxies and clusters of galaxies can only be explained if we assume that there exists another, unknown form of invisible matter, called 'dark matter'.

4. Careful measurements of the expansion history of the universe suggests that 'empty' space actually contains energy. This is called 'dark energy' and is consistent with Einstein's cosmological constant. It explains why spacetime is flat.

5. Detailed measurements (e.g., of the tiny temperature variations in the cosmic background radiation) support a big bang model in which dark energy accounts for sixty-nine per cent of the mass-energy of the universe, dark matter accounts for twenty-six per cent, and ordinary visible matter accounts for just five per cent. The universe is mostly missing.

PART III

WAVE AND PARTICLE

In which waves become particles, and particles become waves. Waves 'collapse', and God plays dice with the universe. And particles gain mass by getting 'dressed'.

9

AN ACT OF
DESPERATION

...what I did can be described simply as an act of desperation....
A theoretical interpretation [of the radiation formula]...had
to be found at any cost, not matter how high.

Max Planck[1]

Einstein's special and general theories of relativity are triumphs of human intellectual endeavour. They are amazingly powerful, used routinely today by scientists and engineers in applications that are at once both esoteric and mundane, touching many aspects of our everyday lives. But the direct effects of relativity are substantially beyond common experience. We will never travel at speeds approaching the ultimate speed, c. We cannot perceive the local curvature of spacetime caused by the Earth (although we can certainly feel the pull of Earth's gravity that results). We cannot reach out and touch dark matter, though it must be all around us. And dark energy—the energy of 'empty' space—though real, is so dilute as to be virtually beyond imagination.

We believe the special and general theories of relativity because we are able to make scientific observations and perform measurements that take us far beyond common experience. But, as we've seen in Part II, these theories serve only further to confuse our understanding of the nature of mass, and thus of material substance itself.

Perhaps this is a good time to turn our gaze away from the large-scale structure of the universe, and look more closely at the small-scale structure of its elementary constituents (well, the elementary constituents of 4.9 per cent of it), in the questing spirit of the early Greek atomists. But let's at least be prepared. Because the truth will prove to be even more shocking.

Let's wind the clock back to December 1900. Although considerable momentum for an atomic interpretation of matter had built since the time of Newton, as we saw in Chapter 4, at the turn of the nineteenth century many physicists were still inclined to be rather stubborn about it. It's perhaps difficult for readers who have lived with the fall-out from the 'atomic age' to understand why perfectly competent scientists should have been so reluctant to embrace atomic ideas, but we must remember that in 1900 there was very little evidence for their existence.

The German physicist Max Planck, for one, was not persuaded. He was something of a reactionary, declaring that the atomic theory was nothing less than a 'dangerous enemy of progress'.[2] Matter is continuous, not atomic, he insisted. He had no doubt that atomic ideas would eventually be abandoned '... in favour of the assumption of continuous matter'.[3]

Planck was a master of classical thermodynamics, a subject that was being pulled inside-out by his arch-rival, Austrian physicist Ludwig Boltzmann. If matter can indeed be reduced to atoms, Boltzmann argued, then thermodynamic quantities such as heat energy and entropy (a measure of the 'disorder' of a physical system) result from statistical averaging over the properties and behaviour of billions and billions of individual self-contained atoms or molecules. To take one example from everyday life, the temperature of the water in your kettle represents a

statistical average of the random motions (and the kinetic energies) of all its molecules. Heating the water simply increases the
speeds of these motions, increasing the temperature.

But statistics have a dark side. They deal with *probabilities*, not
certainties. Classical thermodynamics had by this time established that certain thermodynamic quantities obey certain laws,
with certainty. Perhaps the best-known example is the second
law of thermodynamics. In a closed system, insulated from all
external influences, in any spontaneous change the entropy of
whatever is inside *always* increases. The cables behind your television, set-top box, DVD player, games console, and playbar *will*
become tangled. Disorder will reign. One hundred per cent.

The cocktail glass that in Chapter 3 fell to the floor and shattered (thereby increasing its entropy) follows this second law
behaviour. But Boltzmann's statistics implied that such behaviour is not certain, but simply the most probable outcome for
this kind of system. That the shattered glass might spontaneously reassemble, decreasing its entropy (much to the astonish-
ment of party guests), couldn't be ruled out, although statistically
it is very highly unlikely.

This was too much for Planck. He needed to find a way to show
that Boltzmann's statistical approach was wrong, and in 1897 he
chose the properties of so-called 'black-body' radiation as a bat-
tleground. Heat any object to a high temperature and it will emit
light. We say that the object is 'red hot' or 'white hot'. Increasing
the temperature of the object increases the intensity of the light
emitted and shifts it to a higher range of frequencies (shorter
wavelengths). As it gets hotter, the object glows first red, then
orange-yellow, then bright yellow, then brilliant white.

Theoreticians had simplified the physics somewhat by adopting a model based on the notion of a 'black body', a completely
non-reflecting object that absorbs and emits light radiation

'perfectly'. The intensity of radiation a black body emits is then directly related to the amount of energy it contains.

The properties of black-body radiation could be studied in the laboratory using specialized 'cavities', empty vessels made of porcelain and platinum with almost perfectly absorbing walls. Such cavities could be heated, and the radiation released and trapped inside could be observed with the aid of a small pinhole, a bit like looking into the glowing interior of an industrial furnace. Such studies provided more than just an interesting test of theoretical principles. Cavity radiation was also useful to the German Bureau of Standards as a reference for rating electric lamps.

This must have seemed like a safe choice. Building on earlier work by Austrian physicist Wilhelm Wien and some new experimental results, in October 1900 Planck deduced a 'radiation law' which accounted for all of the available data on the variation in the density of radiation with frequency and temperature. Though the law was elegant, it was in truth no more than a mathematical 'fit' to the data. Planck's challenge now was to find a deeper theoretical interpretation for it.

The law requires two fundamental physical constants, one relating to temperature and a second relating to radiation frequency. The first would gain the label k (or k_B) and become known as *Boltzmann's constant*. The second would eventually gain the label h and become known as *Planck's constant*.

In trying to derive the radiation law from first principles, Planck tried several different approaches. But he found that he was compelled to return to an expression strongly reminiscent of the statistical methods of his rival Boltzmann. The mathematics led him in a direction he really had not wanted to go. He eventually succumbed, in a final act of desperation.

Although the approach Planck took was subtly different from that of Boltzmann, he found that black-body radiation is absorbed and emitted *as though* it is composed of discrete 'atoms', which Planck called *quanta*. Moreover, he found that each quantum of radiation has an energy given by $E = h\nu$, where ν (Greek nu) is the frequency of the radiation. Though much less familiar, this is an expression that is every bit as profound as Einstein's $E = mc^2$.

Planck's own voyage of discovery turned him into a willing and enthusiastic convert to the atomist view, and he presented his new derivation of the radiation law to a regular fortnightly meeting of the German Physical Society on 14 December 1900. This date marks the beginning of the *quantum revolution*.

Planck had used a statistical procedure without giving much thought to its physical significance. If atoms and molecules are real things, something Planck was now ready to accept, then in his own mind the radiation energy itself remained continuous, free to flow uninterrupted back and forth. Planck's own interpretation of $E = h\nu$ was that this reflected the 'quantization' of matter—matter comes in discrete lumps, but the radiation energy it absorbs or emits is continuous.

But there is another possible interpretation. $E = h\nu$ could mean that radiation energy is itself quantized. Could radiation, including light, also be 'lumpy'?

Einstein, for one, was wary of Planck's derivation, figuring that it had involved rather more sleight-of-hand than was satisfactory. By 1905, the evidence for atoms and molecules was becoming overwhelming and this particulate view of matter was now in the ascendancy, as Einstein's own paper on Brownian motion suggested. But did it really make sense to adopt a particulate model for matter and a model for radiation based on continuous waves, as demanded by Maxwell's electromagnetic theory?

Einstein now made a very bold move. In another paper published in his 'miracle year' of 1905, he suggested that $E = h\nu$ should indeed be interpreted to mean that electromagnetic radiation itself consists of 'energy quanta'.[4] This is Einstein's famous 'light-quantum hypothesis'. Two hundred years after Newton, Einstein was ready to return to an atomic theory of light.

He was not proposing to abandon the wave theory completely. There was simply too much experimental evidence for the wave properties of light—phenomena such as light diffraction and interference can only be explained in terms of a wave model. And—lest we forget—the equation $E = h\nu$ connects energy with *frequency*, hardly a characteristic property of 'atoms' in the conventional sense.

Einstein imagined that these two contrasting, or even contradictory, descriptions might eventually be reconciled in a kind of hybrid theory. He believed that what we see and interpret as wave behaviour is actually a kind of statistical behaviour of many individual light-quanta averaged over time.

Whilst special relativity found many advocates in the physics community, the same cannot be said for his light-quantum hypothesis. Most physicists (including Planck) rejected it outright. But this was no idle speculation. Einstein applied the idea to explain some puzzling features associated with a phenomenon known as the *photoelectric effect*, features that were duly borne out in experiments by American physicist Robert Millikan in 1915, earning Einstein the 1921 Nobel Prize for Physics. The idea of light-quanta—'particles of light'—eventually became more acceptable, and the American chemist Gilbert Lewis coined the name *photon* in 1926. It now became possible to draw a few key threads together.

The discovery of the negatively charged electron by Thomson in 1897 implied that atoms, indivisible for more than 2,000 years, now had to be recognized as having some kind of internal structure.* Further secrets were revealed by Rutherford in 1909.

---

* Okay, so strictly speaking what we know today as atoms are not 'atoms' as the Greeks defined them.

Together with his research associates Hans Geiger and Ernest Marsden in Manchester, he had shown that most of the atom's mass is concentrated in a small central nucleus, with the lighter electrons orbiting the nucleus much like the planets orbit the Sun. According to this model, the atom is largely empty space. As a visual image of the internal structure of the atom, Rutherford's planetary model remains compelling to this day.

Rutherford and his colleagues discovered the positively charged proton in 1917. Scientists now understood that a hydrogen atom consists of a single proton in a central nucleus, orbited by a single electron.

The planetary model might be compelling, but it was also understood to be quite impossible. Unlike the Sun and planets, electrons and atomic nuclei carry electrical charge. It was known from Maxwell's theory and countless experiments that electrical charges moving in an electromagnetic field will radiate energy. This is the basis of all radio and television broadcasting. As the 'planetary' electrons lost energy they would slow down, leaving them exposed to the irresistible pull of the positively charged nucleus. Atoms built this way would be inherently unstable. They would collapse in on themselves within about 100 millionth of a second.

And there was another problem. The absorption and emission of radiation by material substances could now be traced to the properties of their atoms and, more specifically, to the electrons within these atoms. Now, when we heat a substance, such as water in a kettle, we find that heat energy is transferred to the water in what appears to be a continuous manner.* The temperature of the water increases gradually, it doesn't suddenly jump

---

* Actually, even the energy of motion of the water molecules in the kettle is quantized—it's just that the individual energy levels are so closely spaced and overlapping that the transfer of energy *appears* to be continuous.

from say 40°C to 70°C. Even though radiation energy comes in lumps, we might nevertheless anticipate that it transfers smoothly and continuously to the electrons inside the atoms of a substance. We might imagine that the electrons 'soak up' the radiation energy as we increase its frequency, until we cross a threshold at which the electrons have too much energy and are ejected.

But this is not the case. As we saw in Chapter 8, it was discovered that atoms absorb or emit light only at a few very discrete frequencies, forming a spectrum consisting of just a few 'lines'. We know that if we pass sunlight through a prism, we will get the familiar rainbow spectrum of colours. But if we look very closely we will see that the harmonious flow from red to violet is actually interrupted by a series of dark lines. These lines appear at frequencies that are absorbed by atoms in the outer layers of the Sun.

The frequencies of the lines in the spectra of individual atoms such as hydrogen appear seemingly at random. However, they are not random. In 1885, the Swiss mathematician Johann Jakob Balmer had studied the measurements of one series of hydrogen emission lines and found them to follow a relatively simple pattern which depends on pairs of integer numbers, such as 1, 2, 3, and so on. This behaviour was extraordinary. The absorption and emission spectra of atoms were providing scientists with a window on the atoms' inner structures. And through this window they glimpsed a baffling regularity.

Balmer's formula was generalized in 1888 by Swedish physicist Johannes Rydberg, who introduced an empirical constant (called the Rydberg constant) into the formula alongside the integer numbers.[5] In themselves, the Balmer and Rydberg formulae are purely empirical—they are mathematical patterns deduced directly from the data. Nobody had any idea how these integer numbers related to the inner structure of the atom.

The orbit of the Earth around the Sun is fixed, to all intents and purposes, with an orbital radius of about 150 million kilometres and an orbital period (the time taken for one complete orbit of the Sun) of a little more than 365 days. In 1913, Danish physicist Niels Bohr wondered if the orbit of an electron around a central proton might be similarly fixed, but instead of only one possible orbit, what if there are several, all of different energy?

In this scenario, absorbing light would pull the electron out of one orbit and shift it out to another, higher-energy orbit, more distant from the nucleus. As the electron returned to the lower-energy inner orbit, it would emit light with a frequency determined by the *difference* in their energies. If this difference is ΔE, where the Greek symbol Δ (delta) denotes 'difference', then the frequency of radiation emitted in the transition from one orbit to the other could be found by rearranging Planck's relation to $\nu = \Delta E/h$.

Bohr found that he needed to impose only one 'quantum' condition, one that introduced an integral number, n, characteristic of each electron orbit.[6] This was sufficient to enable him to derive the generalized Balmer formula and show that the Rydberg constant is simply a collection of other fundamental physical constants, including the mass and charge of the electron, Planck's constant, and the speed of light. Bohr used the values available to him from experiment to calculate the Rydberg constant for hydrogen and got a result that was within six per cent of the measured value, a margin well within experimental uncertainty.

This was yet another example of the use of primarily classical concepts into which a few quantum principles had been shoehorned. But it nevertheless represented a significant step forward. The integer number that characterizes the electron orbits would come to be known as a *quantum number*, and the shifts or 'transitions' between different orbits would be called *quantum jumps*.

The agreement between Bohr's atomic theory and experiment was obviously much more than just coincidence, but the theory begged a lot more questions. The physical impossibility of an electron moving in an orbit around a central proton hadn't been forgotten, but it had been temporarily set aside, a problem to be solved later. Likewise, the 'jumps' between electron orbits had to happen instantaneously if the same problem was to be avoided. Given that the frequency of the radiation emitted by the electron as it jumps from a higher to a lower orbit must match precisely the energy difference (according to $\Delta E = h\nu$), the electron somehow had to 'know' in advance which orbit it was going to end up in.

And, before we get too carried away with excitement, we should acknowledge that Bohr had introduced the quantum number n by *imposing* a quantum condition on its structure. He hadn't answered the question: where do the quantum numbers come from? The outline of an answer came ten years later.

To take this next step we need to think about the energy of a photon according to special relativity. It is reasonably clear how Einstein's equation $E = mc^2$ applies to objects with mass, but how should we now apply it to photons? We know by now that the relativistic mass is given by $m = \gamma m_0$, but countless measurements suggest that photons have zero rest mass, $m_0 = 0$. Photons also travel at the speed of light, so the Lorentz factor γ is infinite. What is infinity times zero? The fact is, I've no idea.

Here's the curious thing. We know for sure that massless photons carry energy. We also know that radiation possesses momentum because we can measure it in the form of radiation 'pressure'. To give one example, the pressure of radiation released by nuclear fusion reactions at the Sun's core balances the gravitational pull at the surface, preventing the Sun from collapsing in on itself.

These facts are reconciled in the expression for the relativistic energy.[7] The energy of a photon with zero rest mass is simply equal to its kinetic energy, $E = pc$, where p is the linear momentum. This might appear to be rather puzzling—isn't kinetic energy supposed to be equal to $\frac{1}{2}mv^2$?—until we realize that this more familiar expression for kinetic energy is an approximation valid for speeds v much less than the speed of light.[8]

We resist the temptation to equate the linear momentum p with mass times velocity (another perfectly valid approximation for speeds v much less than c), allowing us to duck the difficult question about the relativistic mass of the photon. In any case, we will see in Chapter 10 that the cosily familiar classical concept of momentum takes on a completely different guise in quantum mechanics.

It fell to French physicist Prince Louis de Broglie,* the younger son of Victor, fifth duc de Broglie, to put two and two together. The equation $E = pc$ comes from special relativity. The equation $E = h\nu$ comes from Planck's radiation law and Einstein's light-quantum hypothesis. If these are both expressions for the energy of a photon, then why not put them together?

This is what de Broglie did, in 1923. He simply set pc equal to $h\nu$ and, since the frequency of a wave is equal to its speed divided by the wavelength (Greek lambda, λ), he was able to deduce the *de Broglie relation*, $\lambda = h/p$. The wavelength of radiation is inversely proportional to its linear momentum.

The de Broglie relation connects a characteristically wave property—wavelength—on the left with a characteristically particle property—momentum—on the right. This suggests that photons are in some curious sense *both* waves *and* particles. In itself this was a fairly remarkable conclusion, but it was de Broglie's next step that was truly breathtaking. He simply asked himself if this strange behaviour could be somehow universal. What if *electrons* could also be considered to behave like waves?[9]

---

* Pronounced 'de Broy'.

Of course, this is all completely counter-intuitive. It is not our experience that objects with mass behave like waves. But in our daily lives we deal with objects with large masses and therefore large momenta, and the value of Planck's constant is very, very small (it is 6.63×10^{-34} joule-seconds). According to the de Broglie relation, the wavelength of a fast-moving tennis ball is so short as to be well beyond perception.[10]

De Broglie also had a long-standing interest in chamber music, and this now led him to a major breakthrough. Musical notes produced by string or wind instruments are audible manifestations of so-called *standing waves*, vibrational wave patterns which 'fit' within the stopped length of the string or the length of the pipe. A variety of standing wave patterns is possible provided they meet the requirement that they fit between the string's secured ends or the ends of the pipe. This simply means that they must have zero amplitude (they must be 'fastened') at each end. It turns out that this is possible only for patterns which contain an *integral number of half-wavelengths*. In other words, the length of the string or pipe must equal n times $\frac{1}{2}\lambda$, where n is an integer number (1, 2, 3,...) and λ is the wavelength (Figure 10).

If the length of the string or pipe is l, then the longest wavelength standing wave is equal to twice this length, such that $l = \frac{1}{2}\lambda$ ($n = 1$). Such a wave has no points (called 'nodes') where the wave amplitude passes through zero—between the ends. The next wave is characterized by a wavelength equal to l, with one node between the ends, such that $l = \lambda$ ($n = 2$). The wave first rises to a peak, falls back down to zero (the node) and falls further to a trough, before rising back up to zero again. The third has one-and-a-half wavelengths and two nodes, $l = \frac{3}{2}\lambda$ ($n = 3$), and so on.

De Broglie saw that the quantum number n in Bohr's theory could emerge naturally from a model in which an 'electron wave' is confined to a circular orbit around the nucleus. Perhaps, he

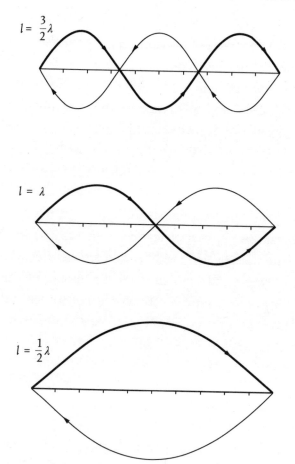

Figure 10. Examples of standing wave patterns. These are characterized by having zero height or amplitude at each end of the stopped string or length of pipe. The arrows give some sense of the direction in which the waves 'travel' although, once established, the waves appear to stand still.

reasoned, the stable electron orbits of Bohr's theory represent standing electron waves, just like the standing waves in strings and pipes which generate musical notes. The same kinds of arguments would apply. For a standing wave to be produced in a circular orbit, the electron wavelengths must fit exactly within the orbit circumference.

Although these ideas were very illuminating, they were nothing more than a loose connection between theoretical concepts. De Broglie had not derived a wave theory of the electron from which the quantum numbers could be expected to emerge naturally. And he had no explanation for the quantum jumps.

The existence of atoms was now accepted, and atoms had been revealed to have an internal structure, with a central nucleus and orbiting electrons. But here, inside the atom, was something unexpected and really rather puzzling. The Greek atomists had argued that matter cannot be divided endlessly to nothing. Two-and-a-half thousand years later the scientists had continued to divide matter, discovering smaller and smaller particles, just as the Greeks might have anticipated. But instead of running up against some kind of ultimate indivisible constituents, the scientists had now discovered particles *that could also be waves*. What on earth was that supposed to mean?

Five things we learned

1. In his pursuit of a description that would refute his atomist rivals, Planck found he had no choice but to embrace a statistical approach and assume that energy could be quantized, as $E = h\nu$.
2. Einstein took this one step further. He argued that this quantization arises because radiation itself comes in 'lumps'. It is composed of discrete light-quanta, which today we call photons.
3. Bohr used Planck's quantum ideas to formulate a model for atomic structure which could explain aspects of the hydrogen atomic spectrum. The structure required curious integral numbers (quantum numbers) and involved instantaneous transitions between different atomic orbits (quantum jumps).

4. De Broglie combined simple expressions for the energy of photons derived from special relativity and quantum theory to deduce the de Broglie relation, $\lambda = h/p$. In doing so he established a connection between a wave-like property (wavelength) and a particle-like property (linear momentum). Particles can also be waves.

5. De Broglie went on to suggest that the quantum numbers in Bohr's theory of the atom might be derived from the need for the 'electron waves' to establish simple standing wave patterns as they orbit the nucleus.

10

THE WAVE EQUATION

> When he had finished, [Pieter] Debye casually remarked that
> this way of talking was rather childish. As a student…he had
> learned that, to deal properly with waves, one had to have a
> wave equation.
>
> *Felix Bloch*[1]

It was called *wave–particle duality*. And, as physicists tried desperately to find ways to come to terms with it, they launched into an unprecedented period of imaginative theorizing. Within a few short years our understanding of the atomic nature of matter would be utterly transformed.

Up to this point, experiment had been firmly in the driving seat. Since Bohr's breakthrough in 1913 ever more detailed studies of atomic spectra had revealed some startling truths. The patterns of absorption and emission lines of the hydrogen atom could be explained through the introduction of the quantum number, n. But, look more closely and what appeared to be one atomic line, with a crisply defined frequency or wavelength, turned out to be two closely spaced lines. Wrap the atoms in a low-intensity electric or magnetic field and some lines would be further split. Quantum numbers proliferated. To n was added k (which, over time, became the *azimuthal* quantum number *l*).* To *n* and *l* was added *m*, the *magnetic* quantum number.

To avoid confusion, the original *n* was renamed the *principal* quantum number, closely linked with the total energies of the electron orbits inside the atom, as Bohr had discovered. Physicists realized that the new quantum numbers reflected constraints on the *geometries* of the different electron orbits and their response to electric and magnetic fields. Relationships between the quantum numbers, such that the value of *n* constrains the possible values for *l*, which in turn constrains the possible values for *m*, told which lines would appear in an atomic spectrum and which were 'forbidden', and therefore missing, spawning an elaborate system of 'selection rules'.[2] Nobody understood where these relationships had come from.

Despite its obvious successes, the fledgling quantum theory of the atom creaked under the strain. Whilst the spectrum of the simplest atom—that of hydrogen—could be accounted for using this rather *ad hoc* system with some degree of confidence, the spectrum of the next simplest atom—helium—could not. The spectra of certain other types of atom, such as sodium and the atoms of rare-earth elements such as lanthanum and cerium, showed 'anomalous' splitting when placed in a magnetic field. The puzzles just kept on coming.

It was becoming increasingly clear that a quantum theory created by shoehorning arbitrary quantum rules into an otherwise

---

* An azimuth is an angular measurement in a spherical co-ordinate system. For example, we can pinpoint a star in the night sky by drawing an imaginary line between the star and us, its observers on Earth. We project this line down to a point on a reference plane, called the horizon (if we're at sea, then the horizon is sea-level). The azimuth is then the angle between this point and a reference direction, such as magnetic north.

classical structure was just not going to work. A completely new 'quantum mechanics' was needed.

The breakthrough, when it came in June 1925, was made by a young German theorist called Werner Heisenberg. He was studying for a doctorate with esteemed physicist Max Born at Göttingen University when he succumbed to a severe bout of hay fever. He left Göttingen to recuperate on the small island of Helgoland, just off Germany's north coast.

Heisenberg had chosen to adopt a firmly empiricist approach, electing to focus his attention on what could be *seen*, rather than what could only be guessed at and speculated on. He reasoned that the secrets of the atom are revealed in atomic spectra, in the precise patterns of frequencies and intensities (or brightness) of individual spectral lines. He now decided that a new quantum mechanics of the atom should deal *only* in these observable quantities, not unobservable electron 'orbits' obliged to obey arbitrary quantum rules.

Free from distractions on Helgoland, he made swift progress. He constructed a rather abstract model consisting of a potentially infinite series of terms, organized into a table with rows and columns, with each term characterized by an amplitude and a frequency, identified with a quantum jump from one orbit to the next in the series. From this table he could work out the intensity of a spectral line resulting from a quantum jump from one orbit to another as the sum of the products of the amplitudes for all possible intermediate jumps.

Whilst this process seemed quite straightforward, Heisenberg became aware of a potential paradox. In conventional arithmetic, if we multiply together two numbers x and y the result (xy) obviously doesn't depend on the order in which we do the multiplication, because x times y gives precisely the same result as y

times x. There's a word for this. The numbers x and y for which $xy = yx$ are said to *commute*.*

But Heisenberg's tables of numbers didn't obey this simple rule. What you got *did* depend on the order in which the intermediate terms were multiplied together. Heisenberg was quite unfamiliar with this kind of result and greatly unsettled by it.

He returned to Göttingen and shared his results—and his concerns—with Born. Born realized that this quirk arises in *matrix* multiplication, a branch of mathematics which deals with the algebra not of single numbers (such as x and y), but of square or rectangular arrays of numbers. Born now worked with his student Pascual Jordan to recast Heisenberg's theory into the language of matrix multiplication. This version of quantum mechanics therefore came to be known as *matrix mechanics*.

In January 1926, Austrian physicist Wolfgang Pauli and English physicist Paul Dirac both independently showed how matrix mechanics could be used to explain key features of the hydrogen atomic emission spectrum, such as the Balmer formula. However they had come about and whatever they were supposed to mean, the matrices of matrix mechanics clearly offered a glimpse of the underlying physics.

But matrix mechanics wasn't to everyone's taste. Older physicists struggled with its mathematical complexity and lack of *anschaulichkeit*, or 'visualizability'. They were therefore quite relieved when, hard on the heels of Heisenberg, Pauli and Dirac (all young upstarts in their early–mid-twenties),† thirty-eight-

---

* English is a remarkable language. A single English word can have many different meanings, revealed only through the context in which it is used. When we say x and y commute we mean $xy = yx$, not that they both travel into work every day.

† Towards the end of 1925 Pauli was twenty-five, Heisenberg twenty-four, and Dirac twenty-three.

year-old Austrian physicist Erwin Schrödinger published details of what appeared to be a completely different theory. This became known as *wave mechanics*.

Intrigued by a throwaway remark in one of Einstein's papers, Schrödinger had acquired a copy of de Broglie's Ph.D. thesis in November 1925. He presented a seminar on de Broglie's work a few days later, attended by physicists at the University of Zurich and the nearby Eidgenossische Technische Hochschule (ETH), the former Zurich Polytechnic where Einstein had completed his graduate studies. In the audience was a young Swiss student named Felix Bloch. Bloch later recalled Schrödinger's presentation, and remarks by Dutch physicist Pieter Debye, who suggested that to deal properly with waves, it is necessary to write down a *wave equation*.[3]

Schrödinger celebrated Christmas 1925 at a villa in the Swiss Alps, taking with him his notes on de Broglie's thesis. By the time he returned on 8 January 1926, he had discovered wave mechanics.

It's possible to follow Schrödinger's reasoning from notebooks he kept at the time. His starting point was the equation of classical wave motion (as Debye had suggested), which describes how an arbitrary wave represented by a *wavefunction*, typically given the symbol ψ (Greek psi), varies in space and time.

Waves can take many different forms, but we are perhaps most familiar with a simple *sine wave*.[4] Schrödinger introduced some quantum behaviour into the classical wave equation by substituting for the wavelength λ using de Broglie's relation ($\lambda = h/p$) and setting the frequency ν equal to E/h.

The resulting wave equation actually represents yet more shoehorning. It is a rather messy fusion of classical wave and classical particle descriptions. It drags an essentially Newtonian concept of mass into the picture, via the (non-relativistic) assumption that the momentum p can be written as mass times velocity.

It is doubtful that anyone would have paid much attention to the result, had it not been for what Schrödinger found he could *do* with it.

A little like Einstein in pursuit of general relativity, Schrödinger quickly reached the limits of his mathematical competence and sought help from a colleague, German mathematician Hermann Weyl. With Weyl's help, Schrödinger discovered that by constraining the properties of a three-dimensional electron wavefunction fitted around a central proton, a very specific pattern of possible solutions of the wave equation emerged entirely naturally.

De Broglie had speculated that the quantum number n might result from the electron forming a standing wave pattern around the proton. This was a good intuition. Schrödinger now demonstrated that a full three-dimensional wave description produced solutions which depend on all three quantum numbers n, l, and m, which emerged 'in the same natural way as the integers specifying the number of nodes in a vibrating string'.[5] Schrödinger also showed that the energies of the various solutions depend on n^2, thereby reproducing the Balmer formula. Make no mistake. These solutions were (and remain to this day) *very* beautiful.

The conclusion was startling. The electron in the hydrogen atom can be described by a wavefunction and the *shapes* of the wavefunctions of different energies are constrained by the electron's quantum nature. Not all shapes are possible. Those that are possible are governed by the values of the quantum numbers, n, l, and m.

Obviously, it's no longer appropriate to call these 'orbits', but we retain the connection with conventional orbits by using the extension 'orbital' instead (see Figure 11). By their very nature such orbitals are distributed through space. The lowest-energy electron orbital has a principal quantum number n equal to

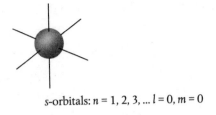

s-orbitals: $n = 1, 2, 3, \ldots l = 0, m = 0$

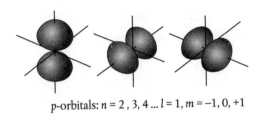

p-orbitals: $n = 2, 3, 4 \ldots l = 1, m = -1, 0, +1$

d-orbitals: $n = 3, 4, 5 \ldots l = 2, m = -2, -1, 0, +1, +2$

Figure 11. Atomic orbitals, illustrated here for the case of an electron in a hydrogen atom. These represent solutions to the Schrödinger wave equation for different combinations of the quantum numbers n, l, and m. The s-orbitals correspond to solutions with $l = 0$ and $m = 0$ and are spherical. The three p-orbitals correspond to solutions with $l = 1$ and $m = -1$, 0, and +1 and are shaped like dumbbells. With increasing energy, the orbital shapes become more and more elaborate, as seen here for the five d-orbitals with $l = 2$ and $m = -2, -1, 0, +1$ and +2.

1 and a value of l equal to zero, and forms a sphere around the nucleus.

So, where could we expect to find the *mass* of the electron in this orbital? Is it also distributed—in some strange way 'smeared'—through the space surrounding the nucleus? But how is this possible? Isn't the electron supposed to be an elementary *particle*?

Born suggested an interpretation that is still taught to science students today. Perhaps, he said, the wavefunction represents a probability for 'finding' the electron at a specific location inside the atom as measured from the nucleus. It's possible to use the wavefunction to construct a probability 'distribution' function, a kind of probability 'cloud'.* For the lowest-energy orbital of the hydrogen atom, this function peaks at a distance measured from the nucleus of 5.29×10^{-11} metres (or 0.053 nanometres), corresponding to the radius of the electron orbit that Bohr had worked out in 1913 in his original atomic theory. It is called the *Bohr radius* and can be calculated from a collection of fundamental physical constants. But the probability distribution function tells us only that the electron has the highest probability of being found at this distance. It can in principle be found *anywhere* within the cloud (see Figure 12). So, where *is* it exactly?

The conceptual nature of the electron had changed, rather dramatically. It was no longer just a 'particle'. It had become a ghost. It could be here, or there, within its orbit. And yet, wherever it was, it had somehow to carry all the electron's mass and electric charge.

And here was something else. The term representing kinetic energy in the Schrödinger wave equation was no longer the familiar $\frac{1}{2}mv^2$ of classical mechanics. It had been replaced by a mathematical 'operator'.[6] Now, we can think of a mathematical operator simply as an instruction to do something to a function, such as 'multiply it', 'divide it', 'take its square root', or 'differentiate it'.

Here's the thing. Just as the result of multiplying two matrices can depend on the order in which we multiply them together,

* No, I don't really know what this means, either.

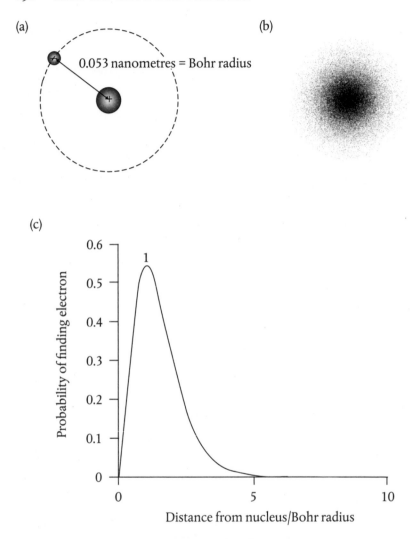

Figure 12. In the Rutherford-Bohr 'planetary' model of the hydrogen atom, a single negatively charged electron occupies a fixed orbit around a nucleus consisting of a single positively charged proton, (a). Quantum mechanics replaces the orbiting electron by an electron 'orbital', within which the electron has a varying probability of being 'found', and the orbital of lowest energy is spherical, (b). The electron can now be found anywhere within this orbital, but has the highest probability of being found at the distance predicted by the old planetary model.

so the result of two mathematical operations can depend on the order in which we perform them. If we multiply a function and then take the square root of the result, we get a different answer compared with taking the square root first and then multiplying the result.[7] Mathematical operations such as these do not commute.

Schrödinger went on to demonstrate that wave mechanics and matrix mechanics are entirely equivalent and give identical results—they express quantum mechanics in two different mathematical 'languages'. Given the choice between Schrödinger's rather beautiful three-dimensional orbitals and Heisenberg's abstract tables of numbers, it's no surprise that the physics community swiftly adopted wave mechanics.

But Heisenberg wasn't quite finished. Both matrix and wave mechanics show that multiplying position (x) by momentum (p) and multiplying p by x produce different answers. This really rather strange property has to have a physical basis. In 1927, Heisenberg realized that it belies an inherent *uncertainty* in the simultaneous measurement of these quantities.

In this context 'uncertainty' simply refers to the precision of an observation or measurement. I might measure a distance x to be *exactly* 1.5 metres, in which case the uncertainty (Δx) is zero. Or I might measure it to be anywhere between 1 and 2 metres, with an average of 1.5 metres and an uncertainty Δx of 0.5 metres either side of the average. Heisenberg determined that if the uncertainty in position is Δx and the uncertainty in momentum is Δp, then the product $\Delta x \Delta p$ *must* be greater than Planck's constant h.* This is Heisenberg's famous *uncertainty principle*.

There is no such principle in the classical mechanics of Newton. In a classical description there is no limit on the precision with

* Today we would write that $\Delta x \Delta p$ must be greater than or equal to $h/4\pi$.

which we can measure the position of an object and its momentum. We are limited only by the precision of our measuring instruments. We assume that a classical object (such as a tennis ball) is precisely in this place, moving at that speed, with that momentum.

Heisenberg's basic premise was that when making measurements on quantum scales, we run up against a fundamental limit. He reasoned that it is simply impossible to make a measurement without *disturbing* the object under study in an essential, and wholly unpredictable, way. At the quantum level, no matter how precise our techniques of measurement, we simply can't get past the fact that they are too 'clumsy'. In this interpretation, quantum mechanics places limits on what is in principle *measureable*.

But Bohr fundamentally disagreed. As Heisenberg had worked out the details of his uncertainty principle, Bohr had pondered on how to make sense of wave–particle duality, and he had arrived at what he believed was a profound, and very different, conclusion. The two argued bitterly.[8]

The contradiction implied by the electron's wave-like and particle-like behaviours was more apparent than real, Bohr decided. We reach for classical wave and particle concepts to describe the results of experiments because these are the only kinds of concepts with which we are familiar from our experiences as human beings living in a classical world. This is a world consisting of particles and waves.

Whatever the 'true' nature of the electron, its behaviour is conditioned by the kinds of experiments we choose to perform. We conclude that in this experiment the electron is a wave. In another kind of experiment, the electron is a particle. These experiments are mutually exclusive. This means that we can ask questions concerning the electron's wave-like properties and we can ask mutually exclusive questions concerning the electron's particle-like properties, but we cannot ask what the electron *really is*.

Bohr suggested that these very different, mutually exclusive behaviours are not contradictory, they are *complementary*. When he heard about the uncertainty principle from Heisenberg, he realized that this places a fundamental limit not on what is measureable, as Heisenberg had come to believe, but rather on what is *knowable*.

The clumsiness argument that Heisenberg had employed suggested that quantum wave-particles actually *do* possess precise properties of position and momentum, and we could in principle measure these if only we had the wit to devise experiments that are subtler, and less clumsy. In contrast, Bohr believed that this has nothing to do with our ingenuity, or lack of it. It has everything to do with the nature of reality at the quantum level. We can't conceive an experiment of greater subtlety because such an experiment is simply *inconceivable*.

Imagine we were somehow able to 'corral' a wave-particle, bundling it up and fixing it in a specific, well-defined region of space, like herding sheep that had wandered all over the hillside into a pen. This would mean that we could, in principle, measure the position of this wave-particle with any level of precision we liked. In theory, this can be done by adding together a large number of waves with different frequencies in what is known as a *superposition*. Waves can be piled on top of each other and added together in ways that particles cannot and, if we choose carefully, we can produce a resulting wave which has a large amplitude in one location in space and a small amplitude everywhere else. This allows us to get a fix on the position, x, of the wave-particle with a high degree of precision, so the uncertainty Δx is very small.

But what about the flip side? The wave-particle also has momentum. Well, that's a bit of a problem. If we express the de Broglie relation in terms of frequency (instead of wavelength), we get $v = pv/h$, where v is the speed of the wave.[9] Now, we localized the wave-particle in one place by combining together lots of waves with different frequencies. This means that we have a large spread of frequencies in the superposition—Δv is large—and this must mean that Δp is large, too. We can measure the position of the wave-particle with high precision, but only at the cost of considerable uncertainty in its momentum.

The converse is also true. If we have a wave-particle with a single frequency, this implies that we can measure this frequency with high precision. Hence we can determine the momentum with similar precision (Δp is very small). But then we can't *localize* the particle; we can't fix it in one place. We can measure the momentum of a quantum wave-particle with high precision, but only at the cost of considerable uncertainty in the particle's position.

The anguished debate between Bohr and Heisenberg wore on. Pauli journeyed to Copenhagen in early June 1927 to act as an impartial referee. With Pauli's support, Bohr and Heisenberg resolved their differences and wounds were healed. But this was no united front, more a rather uneasy alliance. Their perspective became known as the *Copenhagen interpretation* of quantum mechanics.

Schrödinger's wave equation was undoubtedly a triumph, but there was more work still to be done. Despite its successes, the equation did not conform to the demands of special relativity. In fact, Schrödinger himself had discovered that a fully relativistic version produced solutions that disagreed with experiment. Until this problem was resolved, the theory would remain only half-baked.

And there was another puzzle. Bohr had shown that the number of energy levels for an electron in an atom scales with n^2. There is just one level with $n = 1$, four with $n = 2$, nine with $n = 3$, and so on. The different levels with the same n result from the different possible permutations of the quantum numbers l and m and in principle have exactly the same energy unless the atoms are placed in an electric or magnetic field.

We increase the number of protons in the nucleus as we move from hydrogen to helium to lithium, and so on through the entire periodic table. We must then add a balancing number of electrons in order to make the atoms neutral. We might suppose that we simply add one electron to each energy level, or to each orbital. For hydrogen (one proton) we add one electron, filling the $n = 1$ orbital. For helium (two protons) we add two electrons, one in the $n = 1$ orbital and a second in one of the $n = 2$ orbitals.

But if we do this we simply cannot reproduce the pattern exhibited by the atoms as they are arranged in the periodic table. Helium is a 'noble gas'—meaning that it is relatively inert and unreactive. Atoms with electrons hovering on the 'outskirts' of the atom in higher energy orbitals are much more vulnerable and tend to participate actively in chemical reactions. A helium atom with an electron in an outer $n = 2$ orbital could be expected to be a lot more reactive than it is.

In fact, the pattern revealed by the periodic table is consistent with the assumption that each level or orbital can accommodate *two* electrons, not one, such that it scales with $2n^2$. In helium, the two electrons both go into the $n = 1$ orbital, 'closing' this orbital and rendering the atom unreactive and hence 'noble'.[10] Going back to our planetary analogy, whereas there is only one Earth orbiting the Sun at a fixed orbital distance, it seems that inside the atom there can be up to two 'Earths' in each orbit.

Pauli suggested that this must mean that the electron possesses a *fourth* quantum number. For electrons with the same values of n, l, and m to coexist in the same orbital, they must possess different values of this fourth quantum number. This is *Pauli's exclusion principle*.

What could possibly account for this fourth quantum number? There were some clues. A few physicists had earlier suggested that the electron might exhibit a property consistent with a kind of 'self-rotation', acting like a top spinning on its axis, just as the Earth spins as it orbits the Sun. It was Dirac who in 1927 suggested that if the electron can be considered to possess two possible 'spin' orientations then this, perhaps, explains why each orbital can accommodate up to two electrons (Figure 13). The two electrons must be of opposite spin to 'fit' in the same orbital. An orbital can hold a maximum of two electrons provided their spins are *paired*.

Although its interpretation is obscure, we do know from experiment that the electron can line itself up in two different spin directions in a magnetic field. We have learned to think of these possibilities as 'spin-up' and 'spin-down'. If we label the

Spin-up   Spin-down

Figure 13. In 1927, Dirac combined quantum mechanics and Einstein's special theory of relativity to create a fully 'relativistic' quantum theory. Out popped the property of electron spin, imagined as through the negatively charged electron was literally spinning on its axis, thereby generating a small, local magnetic field. Today we think of electron spin simply in terms of its possible orientations—spin-up and spin-down.

electron spin quantum number as s, then this has only one possible value, $s = \frac{1}{2}$. In a magnetic field, the magnetic spin quantum number m_s takes values $+\frac{1}{2}$ (spin-up) and $-\frac{1}{2}$ (spin-down). We re-label the magnetic quantum number m to read m_l (to avoid confusion) and the four quantum numbers that now specify the state of an electron inside the atom are n, l, m_l, and m_s.

Quantum particles with half-integral spins are called *fermions*, named for Italian physicist Enrico Fermi. We're free to make what we like of the term 'spin', but we need to beware. We know that the Earth needs to make one complete rotation around its axis to get back to where it started (and we also know this takes a day). But fermions would have to spin *twice* around their axes in order to get back to where they started. Try to imagine it like this. You can make a Möbius band by taking a length of tape, twisting it once and joining the ends together so the band is continuous and seamless (Figure 14). What you have is a ring of tape with only one 'side' (it doesn't have distinct outside and inside surfaces). Now picture yourself walking along this band. You'll find that, to get back to where you start, you need to walk twice

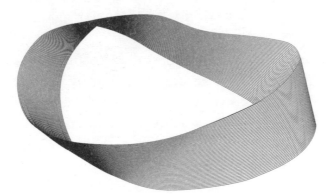

Figure 14. If we start out from any point on a Möbius band, we find that we have to travel twice around it to get back where we started.

around the ring. This just shows that if you persist in pursuing classical analogies in the quantum world, your likely reward will be a bad headache.

Now, electron spin doesn't feature at all in Schrödinger's wave equation, so where could it have come from? Dirac provided the answer towards the end of 1927. In wrestling with a version of the wave equation that would conform to the demands of special relativity, Dirac had represented the wavefunction not as a single function (as Schrödinger had done), but as four functions organized in a square four-by-four array, or matrix. Note that in doing this he had not inadvertently wandered into matrix mechanics. These were still the wavefunctions of wave mechanics, but now organized in a square array.

This was the answer. Two of the four solutions correspond to the two different spin orientations of the electron, and emerge naturally from the mathematics. The other two solutions posed something of a conundrum. What did they represent? Dirac hoped for a simple answer. Still in pursuit of the dream of philosophers, he suggested in 1930 that they might represent the *proton*.

Five things we learned

1. Despite its early successes, Bohr's theory of the atom appeared more and more inadequate as experiment revealed more quantum numbers and demanded a set of elaborate selection rules to explain atomic spectra. A new *quantum mechanics* was needed.
2. Heisenberg provided one version in 1925, based on complex 'tables of numbers' that were later described in terms of matrices. Matrix mechanics found a few early converts but it didn't really take off.

3. Schrödinger's wave mechanics, in contrast, was intuitively simpler and accounted for the quantum numbers in a natural way in terms of three-dimensional 'standing waves'. It also produced rather elegant pictures of electron orbitals. It quickly became the preferred version of quantum mechanics.

4. Heisenberg realized that the curious property of the non-commutation of position and momentum (arising in both wave and matrix mechanics) leads to an inherent *uncertainty* in measurements of these properties. Bohr interpreted this to mean that the wave and particle descriptions are in some sense *complementary*. We can show one or the other behaviour, but not both simultaneously.

5. Schrödinger's wave equation did not meet the demands of special relativity. The fully relativistic wave equation developed by Dirac explained the appearance of a fourth quantum number corresponding to electron spin. But it produced twice as many solutions as needed. Dirac speculated that the other solutions must describe the proton.

11

THE ONLY MYSTERY

We choose to examine a phenomenon which is impossible, absolutely impossible, to explain in any classical way....In reality, it contains the *only* mystery. We cannot make the mystery go away by 'explaining' how it works.

Richard Feynman[1]

Dirac didn't have to wait too long to be disappointed. He had reached for the dream too soon, his proposal roundly criticized on all sides. Among other things, his equation demanded that the masses of the electron and proton should be the same. It was already well known that there is a substantial difference in the masses of these particles, the proton heavier by a factor of almost 2,000.

He finally accepted in 1931 that the other two solutions of his relativistic wave equation must describe a particle with the same mass as the electron. He went on to speculate that these extra solutions imply the existence of a *positive electron*: 'a new kind of particle, unknown to experimental physics, having the same mass and opposite charge to an electron'.[2] American physicist Carl Anderson found evidence for this particle, which he named the *positron*, in cosmic ray experiments in 1932–1933. Dirac had actually predicted *anti-matter*, particles with precisely the same properties as their matter counterparts, but with opposite electrical charge.

And, if that wasn't bad enough, in February 1932 English physicist James Chadwick reported experiments on the nature of the 'radiation' emitted by samples of radioactive beryllium. As this radiation was unaffected by an electric field, it was initially thought to be gamma radiation (very high-energy photons). But Chadwick showed that it rather consists of a stream of neutral particles each with about the same mass as the proton. He had discovered the neutron, which quickly took its place alongside the proton inside the nuclei of atoms heavier than hydrogen.

Although what we now call 'atoms' had been shown to possess an inner structure (and are therefore 'divisible'), their constituents could still be considered to be 'elementary' particles. These could still be viewed as ultimate, indivisible 'lumps' of matter, much as the Greek atomists had envisaged. Yes, particles like electrons happen to have a few properties that had eluded these early philosophers. An electron carries a unit of negative electrical charge. It possesses the rather obscure property of spin, like (but also unlike) a tiny spinning top.

But it seems there is one thing they did get right. It has mass, and we can look up the electron mass in an online catalogue maintained by the *Particle Data Group*, an international collaboration of some 170 physicists which regularly reviews data on elementary particles and produces an annual 'bible' for practitioners.[3]

If we do this, we discover that the mass of the electron is 9.109×10^{-31} kilograms. Now, physicists are just like ordinary human beings. (Trust me on this.) They grow weary of writing numbers with large positive or negative powers of ten, and prefer to find units that allow for a simpler representation. In this case, they associate the electron mass with an energy (using $m = E/c^2$), and write it as 0.511 MeV/c^2. Here MeV stands for mega (million) electron volts and the appearance of '/c^2' in the units serves to remind us of the connection with Einstein's famous equation and indicates that this is a representation of a mass quantity.*

So, let's now have a little fun. Let's imagine that we form a beam of electrons (much like in an old-fashioned television tube) and pass them through a plate in which we've cut two small, closely spaced slits, on the same sort of scale as the wavelength of the electron as given by $\lambda = h/p$. What will we see?

Well, in the particle description we might expect that the electrons in the beam will pass through either one slit or the other. If we place a detector—such as a phosphor screen—on the other side of the slits we would anticipate that we'll see two bright spots marking where the electrons that pass through the slits go on to strike the screen. Each spot will be brightest in the centre, showing where most of the electrons have passed straight through the corresponding slit, becoming more diffuse as we move away from the centre, representing electrons that have scattered on their way through.

But in the wave picture we would expect something rather different. Now in this picture the beam of electrons passes through the slits much like a beam of light, diffracting from each slit and forming an interference pattern of alternating bright and dark fringes, as I described in Chapter 5.

There are two ways in which we might interpret this kind of interference behaviour. We could suppose that the wave nature of the electron is the result of statistical averaging—it arises through some unknown mechanism which affects each individual electron as it passes—as a self-contained elementary particle—through one slit or the other. This is the kind of interpretation that Einstein himself tended to favour.

---

* I will stick with this convention in this book, but you should note that physicists themselves often don't, writing the electron mass as 0.511 MeV. In situations like this the division by c^2 is *implied*.

Or, we might suppose that the wave nature of the electron is an intrinsic behaviour. Each individual electron somehow behaves as a wave distributed through space, passing through both slits simultaneously and interfering with itself. Now this second interpretation would seem to be quite absurd. But there's plenty of experimental evidence to suggest that this is exactly what happens.

Imagine that we now limit the intensity of the electron beam, so that at any moment only one electron passes through the slits. What happens? Well, each electron registers as a bright dot on the phosphor screen, indicating that 'an electron struck here'. As more electrons pass through the slits—one at a time—we see an apparently random scattering of dots at the screen. But then we cross a threshold. As more and more electrons pass through, we begin to see a pattern as the individual dots group together, overlap, and merge. Eventually we see an interference pattern of alternating bright and dark fringes.

These kinds of experiments with electrons have been performed in the laboratory (an example is shown in Figure 15), and there's no escaping the conclusion. Each electron passes through both slits. Its wavefunction on the other side is shaped into high-amplitude peaks and troughs arising from constructive interference and regions of zero-amplitude arising from destructive interference. The high-amplitude peaks and troughs represent regions where there is a high probability of 'finding' the electron: zero-amplitude means zero probability of finding the electron.

When the electron wavefunction reaches the phosphor screen, the electron can in principle be found anywhere. What happens next is utterly bizarre. The interaction with the screen somehow causes the wavefunction to 'collapse'; the location of the resulting dot is then determined by the distribution of probability

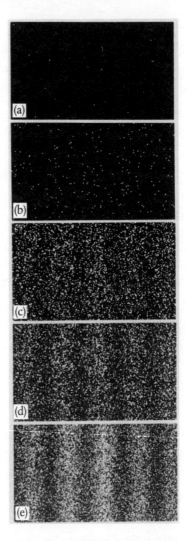

Figure 15. We can observe electrons as they pass, one at a time, through a two-slit apparatus by recording where they strike a piece of photographic film. Each white dot indicates the point where an individual electron is detected. Photographs (a)–(e) show the resulting images when, respectively, 10, 100, 3,000, 20,000, and 70,000 electrons have been detected. The interference pattern becomes more and more visible as the number of detected electrons increases.

derived from the amplitude of the wavefunction.* High probability yields a bright fringe. Low or zero probability yields a dark fringe. Remember that the wavefunction gives only a *probability* for where each electron will end up. It cannot tell us precisely where each electron will be detected.

This all seems totally ridiculous. Why don't we just follow the paths of the electrons as they pass through the slits? Then we would surely see that electrons behave as individual, self-contained lumps of matter which pass through either one slit or the other. And when we do this, particle behaviour is indeed what we get. But the simple truth is that the interaction involved in following the path of the electron collapses its wavefunction prematurely, and so prevents any kind of interference from happening.

We're faced with a choice. If we don't follow the paths of the electrons, we get wave behaviour. If we look to see how we get wave behaviour, we get particle behaviour. These behaviours are complementary, and mutually exclusive. As Bohr insisted, there is no conceivable way of showing both types of behaviour simultaneously.

Einstein was very unhappy about all this. The collapse of the wavefunction and the reliance on probabilities appeared to disturb the delicate balance between cause and effect that had been an unassailable fact (and common experience) for all of human existence. In the world described by classical physics, if we do this, then that will happen. One hundred per cent, no question. But in the world described by quantum physics, if we do this then

* I should point out that this is an assumption. We don't actually know where in the chain of events from detection to human observation the collapse actually occurs. This is the basis for the famous paradox of Schrödinger's cat. Some physicists have suggested that the collapse occurs only when the wavefunction encounters a human consciousness. Others have since suggested that the collapse doesn't actually occur at all, but the universe 'splits' into multiple parallel copies in which different results appear in different universes. Oh, boy!

that will happen with a probability—possibly much less than 100 per cent—that can only be derived from the wavefunction.

Einstein remained stubbornly unconvinced, declaring that 'God does not play dice'.[4]

He was particularly concerned about the collapse of the wavefunction. If a single electron is supposed to be described by a wavefunction distributed over a region of space, where then is the electron supposed to *be* prior to the collapse? Before the act of measurement, the mass (and energy) of the electron is in principle 'everywhere'. What then happens at the moment of the collapse? After the collapse the mass of the electron is localized—it is 'here' and nowhere else. How does the electron get from being 'everywhere' to being 'here' *instantaneously*?

Einstein called it 'spooky action-at-a-distance'.* He was convinced that this violated one of the key postulates of his own special theory of relativity: no object, signal, or influence having physical consequences can travel faster than the speed of light.

Through the late 1920s and early 1930s, Einstein challenged Bohr with a series of ever more ingenious thought experiments, devised as a way of exposing what Einstein believed to be quantum theory's fundamental flaws—its inconsistencies and incompleteness.

Bohr stood firm. He resisted the challenges, each time ably defending the Copenhagen interpretation and in one instance using Einstein's own general theory of relativity against him. But, under Einstein's prosecution, Bohr's case for the defence relied

---

* Because we don't know precisely what happens when the wavefunction collapses, we're probably not justified in declaring that this involves any kind of physical *action* on the particle or particles involved. Einstein may have been hinting that, just as the problem of action-at-a-distance implied by Newton's gravity had been resolved by general relativity, so some kind of further extension of quantum theory was going to be required to fix the problem here.

increasingly on arguments based on clumsiness: an essential and unavoidable disturbance of the system caused by the act of measurement, of the kind for which he had criticized Heisenberg in 1927. Einstein realized that he needed to find a challenge that did not depend directly on the kind of disturbance characteristic of a measurement, thereby completely undermining Bohr's defence.

Together with two young theorists Boris Podolsky and Nathan Rosen, in 1935 Einstein devised the ultimate challenge. Imagine a physical system that emits *two* electrons. We can assume that the physics of the system constrains the two electrons such that they are produced in opposite spin states: one spin-up, the other spin-down.* Of course, we have no idea what the spins actually are until we perform a measurement on one, the other, or both of them.

We'll call the electrons A and B. Suppose electron A shoots off to the left, electron B to the right. We set up our apparatus over on the left. We make a measurement and determine that electron A has a spin-up orientation. This *must* mean that electron B has a spin-down orientation—even though we haven't measured it— as the physics of the system demands this. So far, so good.

Now, in a particle picture, we might assume that the spins of the electrons—whatever they may be—are fixed *at the moment they are produced*. The electrons move apart each with clearly defined spin properties. This is a bit like following the paths of the electrons as they pass through the two slits. Electron A could be either spin-up or spin-down, electron B spin-down or spin-up. But the measurement on electron A merely tells us what spin orientation it had, all along, and by inference what spin orientation electron B must have had, all along. According to this picture, there is no mystery.

---

* This is a variation of the original Einstein–Podolsky–Rosen thought experiment, but it is entirely consistent with their approach.

But it'll come as no surprise to find that quantum mechanics doesn't see it this way at all. The physics of the two electrons is actually described in quantum theory by a *single* wavefunction. The two electrons in such a state are said to be 'entangled'. The wavefunction contains amplitudes from which probabilities can be calculated for all the different possible outcomes of the measurement—spin-up for A/spin-down for B and spin-down for A/spin-up for B.

We don't know what result we will get until we perform the measurement on electron A. At the moment of this measurement, the wavefunction collapses and the probability for one combination (say spin-up for A, spin-down for B) becomes 'real' as the other combination 'disappears'. Prior to the measurement, each individual electron has *no defined spin properties*, as such. The electrons are in a superposition of the different possible combinations.

Here's the rub. Although we might in practice be constrained in terms of laboratory space, we could in principle wait for electron B to travel halfway across the universe before we make our measurement on A. Does this mean that the collapse of the wavefunction must then somehow reach out and influence the spin properties of electron B across this distance? Instantaneously? Einstein, Podolsky, and Rosen (which we'll henceforth abbreviate as EPR) wrote: 'No reasonable definition of reality could be expected to permit this.'[5]

EPR argued that, despite what quantum theory says, it is surely reasonable to assume that when we make a measurement on electron A, this can in no way disturb electron B, which could be halfway across the universe. What we choose to do with electron A cannot affect the spin properties and behaviour of B and hence the outcome of any subsequent measurement we might make on it. Under this assumption, we have no explanation for the sudden change in the spin state of electron B, from 'undetermined' to spin-down.

We might be tempted to retreat to the particle picture and conclude that there is, in fact, no change at all. Electron B *must* have a fixed spin orientation all along. As there is nothing in quantum theory that tells us how the spin states of the electrons are determined at the moment they are produced, EPR concluded that the theory must be incomplete.

Bohr did not accept this. He argued that we cannot get past the complementary nature of the wave-picture and the particle-picture. Irrespective of the apparent puzzles, we just have to accept that this the way nature is. We have to deal with what we can measure, and these things are determined by the way we set up our experiment.

The EPR thought experiment pushed Bohr to drop the clumsiness defence, just as Einstein had intended. But this left Bohr with no alternative but to argue for a position that may, if anything, seem even more 'spooky'. The idea that the properties and behaviour of a quantum wave-particle could somehow be influenced by how we *choose* to set up an apparatus an arbitrarily long distance away is extremely discomforting.

Physicists either accepted Bohr's arguments or didn't much care either way. By the late 1930s, quantum theory was proving to be a very powerful structure and any concerns about what it implied for our interpretation of reality were pushed to the back burner. The debate became less intense, although Einstein remained unconvinced.

But Irish theorist John Bell wasn't prepared to let it go. Any attempt to eliminate the spooky action-at-a-distance implied in the EPR thought experiment involves the introduction of so-called 'hidden variables'. These are hypothetical properties of a quantum system that by definition we can't measure directly (that's why they're 'hidden') but which nevertheless govern those properties that we can measure. If, in the EPR experiment, hidden variables of some kind controlled the spin states of the two electrons such that they are fixed at the moment the electrons are produced, then there would be no need to invoke the collapse of the wavefunction. There would be no instantaneous change, no spooky action-at-a-distance.

Bell realized that if such hidden variables exist, then in certain kinds of EPR-type experiments the hidden variable theory predicts results that disagree with the predictions of quantum theory. It didn't matter that we couldn't be specific about precisely what these hidden variables were supposed to be. Assuming hidden variables of any kind means that the two electrons are imagined to be *locally real*—they move apart as independent entities with defined properties and continue as independent entities until one, the other, or both are detected. But quantum theory demands that the two electrons are 'non-local', described by a single wavefunction. This contradiction is the basis for *Bell's theorem*.[6]

Bell was able to devise a relatively simple direct test. Local hidden variable theories predict experimental results that are constrained by something called *Bell's inequality*—a range of results is possible up to, but not exceeding, a certain maximum limit. Quantum theory predicts results that are not so constrained—they can exceed this maximum, so *violating* the inequality.

Bell published his ideas in 1966. The timing was fortuitous. Sophisticated laser technology, optical instruments, and sensitive detection devices were just becoming available. Within a few years the first practical experiments designed to test Bell's inequality were being carried out. The most widely known of these experiments were performed by French physicist Alain Aspect and his colleagues in the early 1980s, based on the generation and detection of entangled photons (rather than electrons). The results came down firmly in favour of quantum theory.[7]

There are always 'what ifs'. More elaborate hidden variable schemes can be devised to take advantage of 'loopholes' in these kinds of experiments. One by one, ever more ingenious experiments have been carried out to close these loopholes. Today there is really no escaping the conclusion. Reality at the quantum level is non-local. There's no getting around the wave description, the collapse of the wavefunction, and the spooky action-at-a-distance this seems to imply.

But the reality advocated by the proponents of hidden variable theories does not have to be a local reality. The influences of the hidden variables *could* be non-local. How does this help? Well, local hidden variable theories (of the kind that Bell had considered) are constrained by two important assumptions. In the first, we assume that due to the operation of the hidden variables, whatever measurement *result* we get for electron A can in no way affect the result of any simultaneous or subsequent measurement we make on the distant electron B.

In the second, we assume that however we *set up* the apparatus to make the measurement on electron A, this also can in no way affect the result we get for electron B. This is not something to which we would normally give a second thought. As English physicist Anthony Leggett put it: '… nothing in our experience of physics indicates that the [experimental set-up] is either more or less likely to affect the outcome of an experiment than, say, the position of the keys in the experimenter's pocket or the time shown by the clock on the wall'.[8]

We could try to devise a kind of non-local hidden variable theory in which we relax the 'set-up' assumption but keep the 'result' assumption. This would mean that the outcomes of measurements *are* affected by how we choose to set up our apparatus, so we are obliged to accept some kind of unspecified action-at-a-distance that is somewhat spooky. Dropping this assumption means that the behaviour of the wavefunction is somehow governed by the way the apparatus is set up—it somehow 'senses' what's coming and is ready for it. In such a scenario at least the results are in some sense preordained. We get rid of the collapse of the wavefunction and the inherent quantum 'chanciness' that this implies.

This comes down to the rather simple question of whether or not quantum wave-particles like electrons have the properties we measure them to have *before the act of measurement*. Do the 'things-as-they-appear' bear *any* resemblance to the 'things-in-themselves'?

The consequences were worked out in 2003 by Leggett, who was also rather distrustful of the Copenhagen interpretation. Leggett defined a class of what he called 'crypto' non-local hidden variable theories. He found that keeping the result assumption but relaxing the set-up assumption is still insufficient to reproduce all the predictions of quantum theory. Just as Bell had done in 1966, Leggett now derived a relatively simple inequality that could provide a direct experimental test.

Experiments were subsequently performed in 2006 by physicists at the University of Vienna and the Institute for Quantum Optics and Quantum Information. The results were once again pretty unequivocal.[9] Even more unequivocal results using different kinds of quantum states of light were reported by a team from the Universities of Glasgow and Strathclyde in 2010.[10] Quantum theory rules.

The experimental tests of Leggett's inequality demonstrate that we must abandon both the result *and* the set-up assumptions. The properties and behaviour of the distant electron B *are* affected by both the setting we use to measure electron A *and* the result of that measurement. It seems that no matter how hard we try, or how unreasonable the resulting definition of reality, we cannot avoid the collapse of the wavefunction.

We have to accept that the properties we ascribe to quantum particles like electrons, such as mass, energy, frequency, spin, position, and so on, are properties that have no real meaning except in relation to an observation or a measuring device that allows them to be somehow 'projected' into our empirical reality of experience. They are in some sense *secondary*, not primary properties. We can no longer assume that the properties we measure (the 'things-as-they-appear') necessarily reflect or represent the properties of the particles as they really are (the 'things-in-themselves').

At the heart of all this is what charismatic American physicist Richard Feynman declared to be quantum theory's 'only mystery'. Elementary particles such as electrons will behave as individual, localized particles following fixed paths through space. They are 'here' *or* 'there'. But under other circumstances they will behave like non-local waves, distributed through space and seemingly capable of influencing other particles over potentially vast distances. They are 'here' *and* 'there'.

These are now unassailable facts. But one question still nags: where then should we look to find the electron's *mass*?

Five things we learned

1. Individual electrons exhibit wave-interference effects. Their subsequent detection as single, self-contained elementary particles with mass implies that the wavefunction 'collapses'— the probability of finding the electron changes instantaneously from 'anywhere' to 'here'.

2. In seeking to challenge the Copenhagen interpretation of quantum theory, Einstein, Podolsky, and Rosen (EPR) devised an elaborate thought experiment involving entangled particles in which the collapse of the wavefunction would be obliged to happen over potentially vast distances. They argued: 'no reasonably definition of reality could be expected to permit this'.

3. In 1966, Bell further elaborated the EPR thought experiment and developed a simple theorem. 'Fixing' quantum theory using local hidden variables constrains the results of experiments to be within the limits imposed by Bell's inequality. Quantum theory without hidden variables has no such constraints— it predicts results which violate Bell's inequality, providing a direct test.

4. All the tests performed to date have been firmly in favour of quantum theory. Reality is non-local.

5. Leggett devised a further inequality and another subtle test. The properties and behaviour of a distant quantum wave-particle are determined by both the instrument settings we use and the results of measurements we make on its near entangled partner. We cannot assume that the 'things-as-they-appear' necessarily reflect or represent the 'things-in-themselves'.

12

MASS BARE
AND DRESSED

The two talks by Kramers and Lamb stimulated me greatly
and I said to myself well, let's try to calculate that Lamb shift....
I said to myself: I can do that.

Hans Bethe[1]

Aside from the obvious potential to cause headaches, trying
to picture an electron as a wave raises a lot of very difficult,
but nevertheless quite practical, questions. Once again we're
forced to ask what an electron wave is meant to be a wave *in*.
Quantum waves are clearly not like the 'ordinary' waves of classical physics—they're not like ripples on the surface of a pond.
Despite the apparent success of Schrödinger's wave mechanics, it
was quickly realized that even this description told only a part of
the truth.

The structure that emerges from Schrödinger's wave mechanics is based on the idea that physical *properties* such as energy
or linear momentum are 'quantized'. When we add or take away
quanta we are necessarily adding or taking away energy or momentum to or from a system, such as an electron in an atom. This is
all fine, as far as it goes, but it actually doesn't go very far. This
form of quantum mechanics is perfectly satisfactory for describing quantum particles that maintain their integrity in physical

processes, but it can't handle situations in which particles are created or destroyed. In other words, it can't handle quite a lot of physics.

There were also considerable problems with the interpretation of a fully relativistic version of quantum mechanics, which produced solutions with negative energy and (much worse) negative probability. Some physicists in the late 1920s realized that they needed to reach for an alternative structure based on the notion of *quantum fields*.

In this description, it is the quantum field that is 'quantized'. Adding or taking quanta away is then equivalent to adding or taking quantum particles from the field, or creating and destroying particles.* Think of it this way. Conventional quantum mechanics describes the different quantum states of a single particle, like an electron in an atom. Quantum field theory describes the states of the field with different numbers of quantum particles in it.

I'm not sure what pops into your mind at the mention of a 'quantum field', but there's a chance that you might imagine a two-dimensional grid stretching off to the horizon, like a field of wheat, perhaps as painted by Vincent van Gogh. You might even think of 'ripples' in this field as the gentle playing of a breeze over its surface, the swaying ears marking its progress. But quantum fields are three-dimensional. Imagine instead a network or lattice of electrons distributed evenly through space. We could in principle model such a structure using 'point particles' (particles with all their mass concentrated to an infinitesimally small point) and assuming only nearest neighbours interact with each other. Now let's further imagine that we shrink the spacing between these points to zero, or we zoom out so that the spacing between points

---

* For reasons that should be reasonably obvious, the switch to a quantum field description is sometimes called 'second quantization'.

becomes negligible. What we have is then a three-dimensional, continuous 'electron field', the quantum field of the electron.

In this kind of description fields are recognized to be more fundamental than their 'quanta'. Quantum particles become characteristic fluctuations, disturbances, or excitations of the fields. It seemed clear, for example, that the photon must be the quantum of the electromagnetic field, created and destroyed when charged particles interact. A quantum field theory would then be capable of describing the interactions between an electron field (whose quanta are electrons) and the electromagnetic field (whose quanta are photons).

The obvious first step was to take Maxwell's classical theory of the electromagnetic field and try to find a way to introduce quantum-like properties and behaviour in it. If this could be done in a way that satisfied the demands of special relativity, the result would be a quantum version of electrodynamics, called *quantum electrodynamics* or QED.

Heisenberg and Pauli developed a version of just such a quantum field theory in 1929. But there were big problems with it. Maxwell's equations can be solved exactly, meaning that analytical expressions for electrical and magnetic properties and behaviour can be derived by simply solving the equations using the rules of conventional (though somewhat abstract) mathematics. These properties and behaviour can then be calculated in a relatively straightforward manner directly from the solutions. The early version of QED was not so kind, however. It did not yield exact solutions.

The theorists had no choice but to resort to an approximation. When faced with an insoluble mathematical problem, one approach is to approximate this as a problem that *can* be solved exactly, to which is added a series of 'perturbation' terms—literally terms which 'perturb' the simple solution. These added terms form a power series, increasing in the power (squared, cubed, fourth-power, etc.) of some relevant quantity, such as energy.

For QED, the starting point is a quantum field theory expression involving no interaction between the electron and the electromagnetic field, which can be solved exactly. In principle, each added perturbation term should then provide a smaller and smaller correction to the starting point, getting us closer and closer to the actual result.[2] The accuracy of the final result would then depend simply on the number of terms included in the calculation (and the patience of the theorist).

Note my use of the words 'in principle' and 'should'. It was here that Heisenberg and Pauli ran into problems. The second-order (squared) term in the series should have provided a small correction to the zero-interaction starting point. Instead it yielded a correction that mushroomed to infinity. This made no physical sense—infinity is a mathematical abstraction and does not exist in nature.

The problem was traced to the 'self-energy' of the electron. The infinity results from an electron interacting with its own self-generated electromagnetic field, relentlessly absorbing and emitting photons of all energies in a veritable blizzard. Obviously, in reality electrons do not possess infinite energy. Nature has clearly found a way to limit these interactions. But there was no obvious way to limit their mathematical expression in the early versions of QED. No further progress could be made until this problem was either eliminated or resolved.

In the early 1930s, it seemed as though quantum field theory might be a dead end. But some physicists realized that, if the problems could be solved, the theory offered a very different way to understand how forces between particles might actually work. Suppose that two electrons are 'bounced' off each other (the technical term is 'scattered'). We know that like charges repel, much like the north poles of two bar magnets. So we imagine that as the two electrons approach each other in space and in time they feel the mutually repulsive electrostatic force generated by their negative charges.

But how? We might speculate that each moving electron generates an electromagnetic field and the mutual repulsion is felt in the space where these two fields overlap, much like we feel the repulsion of the north poles of two bar magnets in the space between them. But in the quantum domain, fields are associated with particles, and interacting fields are therefore associated with 'particles of interaction'. In 1932, German physicist Hans Bethe and Enrico Fermi suggested that the experience of this force of repulsion is the result of the *exchange of a photon* between the two electrons (see Figure 16).*

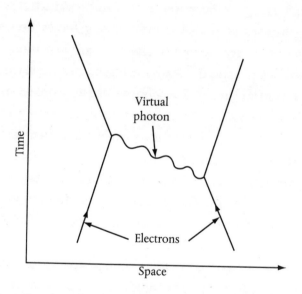

Figure 16. A Feynman diagram representing the interaction between two electrons as described by quantum electrodynamics. The electromagnetic force of repulsion between the two negatively charged electrons involves the exchange of a virtual photon at the point of closest approach. The photon is 'virtual' as it is not visible during the interaction.

---

* This means that when we try to push the north poles of two bar magnets together, the resistance you feel is caused by invisible photons passing back and forth between them.

The exchanged photon carries momentum from one electron to the other, thereby changing the momentum of both. The result is recoil, with both electrons changing speed and direction and moving apart.

The exchanged photon is a 'virtual' photon, because it is transmitted directly between the two electrons and we don't actually see it pass from one to the other. According to this interpretation, the photon is no longer simply the quantum particle of light. It has become the 'carrier' of the *electromagnetic force*.

Now here's a thing. Matter particles (such as protons, neutrons, and electrons) are all *fermions*, with half-integral spin quantum numbers ($s = \frac{1}{2}$). Fermions obey Pauli's exclusion principle. But the particles responsible for carrying forces between the matter particles are different. They are all *bosons* (named for Indian physicist Satyendra Nath Bose), characterized by integral spin quantum numbers. The photon, for example, has a spin quantum number $s = 1$.

Bosons are not subject to Pauli's exclusion principle: they *can* possess the same quantum numbers and can 'condense' into a single quantum state. An example of such 'bose condensation' is laser light. It is impossible to achieve similar condensation with a beam of electrons—to produce an 'electron laser'. Pauli's exclusion principle forbids this.

Despite the attractions of QED and this explanation in terms of matter and force particles, no progress could be made until the problems of quantum field theory were resolved. Over the next fifteen years or so—interrupted by the turmoil of world war and the development of atomic weapons—the theory spilled into crisis.

A faint light at the end of the long dark tunnel of incomprehension was eventually spied in early June 1947, at a small, invitation-only conference held at the Ram's Head Inn, a small clapboard hotel and inn on thinly populated Shelter Island, at the eastern end of New York's Long Island. At this conference, leading theorists heard details of two experiments which, at first sight, seemed only to make things worse.

Columbia University physicist Willis Lamb, a former student of J. Robert Oppenheimer (the 'father' of the atom bomb) reported on some disturbing new results on the spectrum of the hydrogen atom. He had focused on the behaviour of two energy levels which share the same principal quantum number $n = 2$, but possess values of the azimuthal quantum number, l, of 0 and 1. These correspond to electron orbitals with different shapes. Orbitals with $l = 0$ form spheres around the central nucleus. Orbitals with $l = 1$ form lobes pointing in opposite directions shaped rather like dumbbells.

In Dirac's relativistic quantum theory, the energies of these different orbitals is determined only by the value of n (as it was in Bohr's theory and Schrödinger's wave mechanics). Despite their different shapes, in the absence of an external electric or magnetic field these orbitals were expected to have precisely the same energy, producing a single line in the atomic spectrum.

But, working with graduate student Robert Retherford, Lamb had found that there are in fact *two* lines. One of the levels is shifted very slightly in energy relative to the other, a phenomenon that quickly became known as the *Lamb shift*. The physicists gathered on Shelter Island now heard the full details of Lamb's latest experimental results.

Galician-born physicist Isidor Rabi then stood to deliver a second experimental challenge. He reported the results of experiments conducted by his Columbia University students John Nafe and Edward Nelson and his Columbia colleagues Polykarp Kusch and H.M. Foley. These were concerned with measurements of the 'g-factor' of the electron, a fundamental physical quantity which governs the extent to which an electron interacts with a magnetic field. According to Dirac's theory, the g-factor should have a value exactly equal to 2. Rabi now reported that careful measurements had revealed that the g-factor is actually slightly larger, more like 2.00244.

Oppenheimer felt sure that these small discrepancies between predictions based on Dirac's theory and measurement had everything to do with quantum electrodynamic effects. The differences were small, but very real. And they were certainly not infinite.

Clues to the eventual solution were revealed on the second day of the conference. Dutch theorist Hendrik Kramers gave a short lecture summarizing his recent work on a classical theory of the electron, outlining a new way of thinking about the electron's mass in an electromagnetic field. Suppose the self-energy of the electron (which plagues the second-order perturbation term in QED) appears as an *additional contribution to the electron's mass*. The mass that we observe in experiments would then include this contribution. In other words, the observed mass of the electron is actually equal to its intrinsic mass plus an 'electromagnetic mass' arising from the electron's interaction with its own self-generated electromagnetic field.

Physicists use some other terms to describe what's going on here that are quite evocative. An electron stripped of its covering electromagnetic field is a 'naked' or 'bare' electron, and its mass is sometimes referred to as the 'bare mass'. As an electron can never be separated from its electromagnetic field, this is a purely hypothetical quantity. The mass that the physicists have to deal with is rather the observed, or 'dressed mass', the mass that the electron possesses by virtue of its covering electromagnetic field.

After the conference, Bethe returned to New York and picked up a train to Schenectady, where he was engaged as a part-time research consultant to General Electric. Like many of his contemporaries, he had been thinking deeply about the Lamb shift, and the discussions during the conference now prompted him to attempt a calculation. As he sat on the train he played around with the equations.

The existing theories of QED predicted an infinite Lamb shift, a consequence of the self-interaction of the electron with its own electromagnetic field. Bethe now followed Kramers' suggestion and identified the troublesome term in the perturbation series with an electromagnetic *mass* effect. His logic runs something like this. The electron in a specific energy level in a hydrogen atom has a certain self-energy which corresponds to the electron's electromagnetic mass. (Let's overlook for the moment that QED says this is infinite.) The electron's total energy is then given by the energy it possesses by virtue of the specific atomic orbital it occupies, plus the self-energy. In mass terms, it has a total mass given by the bare mass associated with the orbital, plus the electromagnetic mass.

A free electron—completely removed from the hydrogen atom—has a total energy given by its kinetic energy of motion, plus the self-energy. Or, it has a total mass given by the bare mass associated with a free electron, plus the electromagnetic mass. In both situations the electromagnetic mass is the same. So why don't we get rid of it by subtracting one from the other?

It sounds as though subtracting infinity from infinity should yield a nonsensical answer, but Bethe now found that in a non-relativistic version of QED this subtraction produced a result that, though still rather troublesome, was much better behaved. Aided by some intelligent guesswork, he was able to obtain a theoretical estimate for the energies of the two energy levels of the hydrogen atom affected and so estimate the size of the Lamb shift.

He obtained a result just four per cent larger than the experimental value that Lamb had reported. Oppenheimer had been right. The shift is a purely quantum electrodynamic effect. Bethe figured that in a fully relativistic QED this 'mass renormalization'

(a)

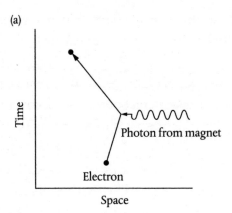

(b)

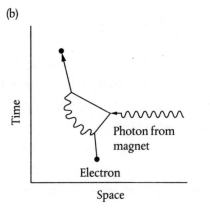

(c)

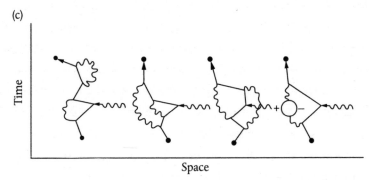

Figure 17. Feynman diagrams representing the interaction of an electron with a photon from a magnet. If the electron is assumed to absorb the photon, as in (a), this would predict a g-factor for the electron of exactly 2, as given by Dirac's theory. But if we include electron self-interaction, depicted as the emission and re-absorption of a virtual photon, (b), this leads to a slight increase in the prediction for the g-factor. Further 'higher-order' processes (c) involving emission and re-absorption of two virtual photons and the spontaneous creation and annihilation of an electron-positron pair are much less probable but add further small contributions.

procedure would eliminate the problem completely and give a physically realistic answer.

So, what causes the Lamb shift? Picture an electron inside a hydrogen atom. All that relentless activity involved in absorbing and emitting virtual photons as the electron interacts with its own electromagnetic field results in a slight 'wobble' in the electron's motion. This adds to its orbital and spin motions around the central nucleus, 'smearing' its probability. The effect is most noticeable when the electron occupies an orbital that, on balance, keeps it closer to the nucleus. Consequently, the energy of the spherical ($l = 0$) orbital is raised slightly compared with the energy of the dumbbell-shaped ($l = 1$) orbitals, since on average it spends more time closer to the nucleus in the former.

Accounting for the discrepancy in the electron g-factor required a fully relativistic version of QED. This was eventually devised by rival American physicists Richard Feynman and Julian Schwinger, and independently by Japanese theorist Sin-Itiro Tomonaga. English physicist Freeman Dyson subsequently demonstrated that their different approaches were entirely equivalent.

The starting point for the evaluation of the g-factor is the interaction between an electron and a virtual photon emitted (for example) from a magnet. If we base our calculation of the g-factor on this single interaction, we will get the same result predicted by Dirac's theory—a g-factor of exactly 2.

But there are other ways this process can happen. A virtual photon can be emitted and absorbed by the same electron, representing an electron interacting with its own electromagnetic field. If we include this process, we get a g-factor very slightly larger than 2.

We can go on. The self-interaction can involve two virtual photons. It can involve more complicated processes, for example, in which a single virtual photon spontaneously creates an electron-positron pair, which then mutually annihilate to form another photon which is then subsequently absorbed. The possibilities are endless, although the more elaborate the process the lower its probability and the smaller the correction to the end result (Figure 17).

Physicists realized that in the microscopic world of quantum fields it is important to change the way we think. For sure, there can be no circumstances under which energy is not conserved, but that doesn't prevent an awful lot of strange things from happening. Heisenberg's uncertainty principle is not limited to position and momentum. It applies to other pairs of physical properties, called *conjugate* properties, such as energy and time. Just as for position and momentum, we write that ΔE times Δt cannot be smaller than $h/4\pi$, where ΔE is the uncertainty in energy and Δt is the uncertainty in the rate of change of energy with time.

So, imagine we now create a perfect vacuum, completely insulated from the external world. If we temporarily forget everything we know about dark energy, we might be tempted to argue that there is 'nothing' at all in this vacuum. What does this imply in terms of quantum field theory and the uncertainty principle? It implies that the energy of an electromagnetic field (or any other field for that matter) in the vacuum is zero. It also implies that the rate of change of the energy of this field is zero, too. But the uncertainty principle denies that we can know precisely the energy of an electromagnetic field and its rate of change. They *can't* both be exactly zero.

Now, the uncertainty principle doesn't expressly forbid 'borrowing' the energy required to create a virtual photon or electron–positron pair literally out of nothing, so long as it is 'given back' within a time that conforms to the demands of the uncertainty principle. The larger the energy borrowed, the quicker is has to be given back. In 1958, American theorist Murray Gell-Mann declared what he called the 'Totalitarian Principle', which states: 'Everything not forbidden is compulsory.'[3] We can interpret this to mean that if it isn't forbidden by the uncertainty principle, then it *must* happen.

As a consequence of the uncertainty principle, the vacuum field experiences random quantum fluctuations, like turbulent waves in a restless ocean. Now, fluctuations in a quantum field are equivalent to particles, and it's tempting to imagine all sorts of elementary particles constantly popping in and out of existence in the vacuum. But this would be a little misleading. If the quantum field is an orchestra, then the random fluctuations are a cacophony of disconnected and discordant notes which combine to make 'noise', as though the orchestra is tuning up. These are the 'virtual' particles. Occasionally, and quite randomly, a few pure notes can be heard. These are the fundamental vibrations of the field that we identify with elementary particles.

The fluctuations average out to zero, both in terms of energy and its rate of change, but they can nevertheless be non-zero at individual points in spacetime. 'Empty' space is in fact a chaos of wildly fluctuating quantum fields and virtual particles.*

---

* Ha! Could this be the answer to the mystery of dark energy? If the vacuum is fizzing with virtual particles, surely this contributes to its energy? Alas, no. As the

How can this possibly be right? Well, we do have some evidence in the form of something called the *Casimir Effect*, predicted by Dutch theorist Hendrik Casimir in 1948.

Take two small metal plates or mirrors and place them side by side a few millionths of a metre apart in a vacuum, insulated from any external electromagnetic fields. There is no force between these plates, aside from an utterly insignificant gravitational attraction between them which, for the purposes of this experiment, we can safely ignore.

And yet, although there can be no force between them, the plates are actually pushed very slightly together. What's happening? The narrow space between the plates forms a cavity, limiting the number of virtual photons that can persist there.* The density of virtual photons between the plates is then lower than the density of virtual photons elsewhere. The end result is that the plates experience a kind of virtual radiation pressure; the higher density of virtual photons on the outsides of the plates pushes them closer together. This effect was first measured by physicist Steven Lamoreaux at Los Alamos National Laboratory in 1996, who obtained a result within five per cent of the theoretical prediction. Subsequent experiments have reduced the difference to one per cent.

---

uncertainty principle 'allows' near-infinite energy fluctuations so long as these occur within an infinitesimal amount of time, quantum field theory predicts that the energy of the vacuum should be essentially infinite. More modest estimates based on some rather arbitrary assumptions predict a vacuum energy of 10^{105} joules per cubic centimetre. Recall from Chapter 8 that the mass density of 'empty' spacetime is 8.6×10^{-30} grams per cubic centimetre. We use $E = mc^2$ to convert this into an energy density of 5.3×10^{-16} joules per cubic centimetre. So, the prediction is out by a factor of 10^{121}. Oops. That's possibly the worst prediction in the entire history of science.

* A bit like de Broglie's musical notes, only virtual photons with standing waves that 'fit' between the plates will survive interference.

Such interactions affect the size of the electron's g-factor. Interactions between the electron and all the virtual photons that 'dress' it add an electromagnetic mass, but they also cause a small amount of this mass to be carried away. The magnitude of the electron's charge is unchanged. This subtly affects the way that the dressed electron interacts with a magnetic field.

Not everyone was comfortable with mass renormalization (Dirac—the mathematical purist—thought it was 'ugly'), but there was no denying the awesome power of a fully relativistic QED. The g-factor for the electron is predicted by the theory to have the value 2.00231930476. The comparable experimental value is 2.00231930482. 'This accuracy', wrote Feynman, 'is equivalent to measuring the distance from Los Angeles to New York, a distance of over 3,000 miles, to within the width of a human hair.'

It's time to take stock. With the development of QED and the mathematical sleight-of-hand involved in mass renormalization, our understanding of the elementary constituents of matter has taken a further troubling turn. Particles—those ultimate, indivisible bits of 'stuff' beloved of the early Greek atomists—have been replaced by quantum fields. What we still tend to think of as particles are no more than characteristic disturbances of these fields. Matter has been reduced to ghosts and phantoms.

And what of mass? As we saw in Chapter 11, the mass of an electron is measured to be 0.511 MeV/c^2. There was a time when we might have been tempted to think of this in much the same way that Newton did, as *quantitas materiae*, the amount of matter in an electron. No more.

We now know that the mass of an electron is a composite. It consists of a bare mass, the hypothetical mass an electron would possess if it could ever be separated from its own, self-generated electromagnetic field. To this we must add electromagnetic mass, created by the *energy* of the countless interactions between the electron and its electromagnetic field, absorbing and emitting the virtual photons which 'dress' it.

Five things we learned

1. The complementary wave-like and particle-like behaviour exhibited by electrons encouraged physicists to develop a quantum version of Maxwell's classical theory of electrodynamics. The result is quantum electrodynamics, or QED.

2. It was discovered in early versions of QED that the equations could not be solved exactly. Approximate solutions, involving the use of a perturbation series, were sought but immediately encountered problems when the second-order term in the series was found to be infinite. This is a problem with the mathematics that nature has obviously found a way to avoid.

3. The experimental determination of the magnitude of the Lamb shift and a g-factor for the electron slightly greater than 2 pointed the way to a solution. The electron has a 'bare' mass and an 'electromagnetic mass', the latter resulting from interactions with its own self-generated electromagnetic field which 'dresses' the electron. The troublesome infinite term is a direct result of these self-interactions.

4. Bethe found that the infinite term could now be removed by subtracting the equation for a free electron from the equation for an electron bound in a hydrogen atom, in a process called *mass renormalization.*

5. Mass took another step towards obscurity. The mass of an electron is, in part, derived from the *energy* of the virtual photons that dress it.

PART IV

FIELD AND FORCE

In which elementary particles become fundamental vibrations or fluctuations of quantum fields; protons and neutrons become composites made of quarks, tethered by gluons; mass becomes a behaviour rather than a property; and we realize that there is no such thing as mass, there is only the energy content of quantum fields.

13

THE SYMMETRIES
OF NATURE

That is not a sufficient excuse.

Wolfgang Pauli[1]

Even though physicists struggled to come to terms with what quantum theory was telling them about the nature of the material world, there could be no denying the theory's essential correctness. Today the theory continues to make predictions that defy comprehension, only for these same predictions to be rigorously upheld by experiment.

But the quantum theory we have surveyed in Chapters 9–12 is primarily concerned with the properties and behaviour of electrons and the electromagnetic field. Now, it's true to say that the behaviour of electrons which surround the nuclei of atoms underpins a lot of physics and pretty much all of chemistry and molecular biology. Yet we know that there is much more to the inner structure of atoms than this. With the problems of quantum electrodynamics (QED) now happily resolved, the attentions of physicists inevitably turned inwards, towards the structure of the atomic nucleus itself.

QED was extraordinarily successful, and it seemed to physicists that they now had a recipe that would allow them to establish theories of other forces at work inside the atom. By the early

1950s, it was understood that there were three of these. Electro-magnetism is the force that holds electrically charged nuclei and electrons together. The other two forces work on protons and neutrons *inside* the atomic nucleus.

The first of these is called the weak nuclear force. As the name implies, it is much weaker than the second kind of force which—for rather obvious reasons—is called the strong nuclear force. The weak nuclear force manifests itself in the form of certain types of radioactivity, such as beta-radioactive decay. This involves a neutron spontaneously converting into a proton, accompanied by the ejection of a fast-moving electron. We can write this as $n \rightarrow p^+ + e^-$, where n represents a neutron, p^+ a positively charged proton, and e^- a negatively charged electron.

In fact, a free neutron is inherently radioactive and unstable. It has a half-life of about 610 seconds (a little over ten minutes), meaning that within this time we can expect that half of some initial number of neutrons will decay into protons (or, alternatively, there's a fifty per cent chance that each neutron will have decayed). Neutrons become stabilized when they are bound together with protons in atomic nuclei, so most nuclei are not radioactive. There are a few exceptions, however. For example, the nucleus of an isotope of potassium with nineteen protons and twenty-one neutrons has a little too much energy for its own good. It undergoes beta-radioactive decay with a half-life of about 1.3 billion years, and is the most significant source of naturally occurring radioactivity in all animals (including humans).

The involvement of an electron in beta-radioactivity suggests that the weak nuclear force operates on electrons, too, although it is very different from electromagnetism. For this reason I'll drop the name 'nuclear' and call it the 'weak force'. There's actually a little more to this force. Careful analysis of the energies of the particles we start with in beta-radioactive decay compared with the particles we end up with gives a small discrepancy. In 1930, Pauli suggested that another kind of uncharged particle must also be ejected along with the electron, responsible for carrying away some of the energy of the transformation. The mysterious particle was called the *neutrino* (Italian for 'small neutral one') and was discovered in 1956. Like the photon, the neutrino is electrically neutral and for a long time was thought to be massless. Neutrinos are now believed to possess very small masses.

So, to keep the energy bookkeeping above-board, we now write: $n \rightarrow p^+ + e^- + \bar{v}$ where $\bar{v}$ (pronounced 'nu-bar') represents an anti-neutrino, the anti-particle of the neutrino.

Now, if QED is a specific form of quantum field theory, the key question the theorists needed to ask themselves in the early 1950s was this: what kind of quantum field theories do we need to develop to describe the weak and strong forces acting on protons and neutrons?

As they scrambled for clues, theorists reached for what is arguably one of the greatest discoveries in all of physics. It provides us with a deep connection between critically important laws of *conservation*—of mass-energy, linear, and angular momentum* (and many other things besides, as we will see)—and basic *symmetries* in nature.

In 1915, German mathematician Emmy Noether deduced that the origin of conservation laws can be traced to the behaviour of physical systems in relation to certain so-called *continuous symmetries*. Now, we tend to think of symmetry in terms of things like rotations or mirror reflections. In these situations, a symmetry transformation is equivalent to the act of rotating the object

---

* Angular momentum is simply the rotational equivalent of linear momentum. Think about your experiences as a child playing on a merry-go-round or carousel. The faster you whirl around in a circle, the greater your angular momentum.

around a centre or axis of symmetry, or reflecting an object as though in a mirror.

We claim that an object is symmetrical if it looks the same following such a transformation. So, we would say that a diamond symbol, such as ◆, is symmetrical to 180° rotation around its long axis (top to bottom) or when reflected through its long axis, the point on the left mirroring the point on the right.

These are examples of *discrete* symmetry transformations. They involve an instantaneous 'flipping' from one perspective to another, such as left-to-right or top-to-bottom. But the kinds of symmetry transformations identified with conservation laws are very different. They involve gradual changes, such as continuous rotation in a circle. Rotate a perfect circle through a small angle measured from its centre and the circle obviously appears unchanged. We conclude that the circle is symmetric to continuous rotational transformations around the centre. We find we can't do the same with a square or a diamond. A square is not continuously symmetric—it is instead symmetric to discrete rotations through 90°, a diamond to discrete rotations through 180° (see Figure 18).

Noether found that changes in the *energy* of a physical system are symmetric to continuous changes in *time*. In other words, the mathematical laws which describe the energy of a system now will be exactly the same a short time later. This means that these relationships do not change with time, which is just as well. Laws that broke down from one moment to the next could hardly be considered as such. We expect such laws to be, if not eternal, then at least the same yesterday, today, and tomorrow. This relationship between energy and time is the reason these are considered to be conjugate properties, their magnitudes governed by an uncertainty relation in quantum theory.

The laws describing changes in linear momentum are symmetric to continuous changes in *position*. The laws do not depend

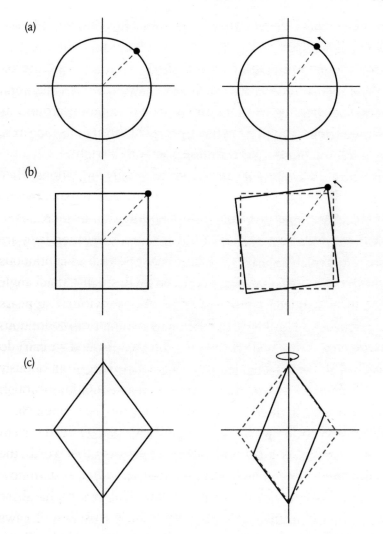

Figure 18. Continuous symmetry transformations involve small, incremental changes to a continuous variable, such as a distance or an angle. When we rotate a circle through a small angle, the circle appears unchanged (or 'invariant') and we say that it is symmetric to such transformations, (a). In contrast, a square is not symmetric in this same sense. It is, instead, symmetric to *discrete* rotations through 90°, (b). A diamond is symmetric to 180° rotations around its long axis, (c).

on where the system is. They are the same here, there, and everywhere, which is why position (in space) and linear momentum are also conjugate properties, governed by an uncertainty relation.

For angular momentum, defined as motion in a circle at a constant speed, the equations are symmetric to continuous changes in the *angle* measured from the centre of the rotation to points on the circumference. And the answer to your next question is yes, there are uncertainty relations that govern the angular momentum of quantum systems.

Once the connection had been established, the logic of Noether's theorem could be turned on its head. Suppose there is a physical quantity which appears to be conserved but for which the laws governing its behaviour have yet to be worked out. If the physical quantity is indeed conserved, then the laws—whatever they are—must be symmetric in relation to some specific continuous symmetry. If we can discover what this symmetry is, then we are well on the way to figuring out the mathematical form of the laws themselves. Physicists found that they could use Noether's theorem to help them find a short cut to a theory, narrowing the search and avoiding a lot of unhelpful speculation.

The symmetry involved in QED is represented by something called the U(1) symmetry group, the unitary group of transformations of one complex variable, also known as the 'circle group'. A symmetry 'group' is a bit like a scoreboard. It represents all the different transformations for which an object is symmetrical. The number in brackets refers to the number of 'dimensions' associated with it—in this case, just 1. These are not the familiar dimensions of spacetime. They are abstract mathematical dimensions associated with the properties of the different transformations. It really doesn't matter too much what the symbol stands for, but we can think of U(1) as describing transformations synonymous with continuous rotations in a circle.[2] This symmetry ties the electron and the electromagnetic field together in an intimate embrace. The upshot is that, as a direct result, *electric charge is conserved*.

The charge of the electron is preserved in all interactions with photons, so the photon is not required to carry a charge of its own. Also, the electromagnetic field acts over long ranges (though the strengths of these interactions fall off with distance). This means that the electromagnetic force can be carried quite happily by neutral, massless particles, able to travel long distances at the speed of light, which we call photons.

Although they are rather abstract, the dimensions of the symmetry groups have important consequences which are reflected in the properties and behaviours of particles and forces in our physical world. At this stage it's useful just to note that a U(1) quantum field theory describes a force carried by a single force particle, the photon, acting on electrically charged matter particles, such as protons and electrons.

This is all very fine for electromagnetism, but the strong and weak forces clearly act on both positively charged protons and neutral neutrons. In beta-decay, a neutron transforms into a positively charged proton, so although the charge is balanced by the emitted electron, the charge on the particle acted on by the weak force is *not* conserved.

It is also pretty obvious that the strong and weak forces must be very short-range forces—they appear to act only within the confines of the nucleus, distances of the order of a femtometre (10^{-15} metres), in stark contrast to electromagnetism. In 1935, Japanese physicist Hideki Yukawa suggested that the carriers of short-range forces should be 'heavy' particles. Such force carriers would move rather sluggishly between the matter particles, at speeds much less than light.[3] Yukawa went on to predict that the carriers of the strong force ought to have masses of the order of ~100 MeV/c^2.[4]

There was no lack of interest in the weak force, as we will soon see, but the strong force appeared to hold the key to understanding the physics of the nucleus itself. As physicists began to think about the formulation of quantum field theories to describe the strong force, they asked themselves: what physical property is conserved in strong force interactions? The answer wasn't all that obvious.

The proton and neutron have very similar masses—938.3 and 939.6 MeV/c^2, according to the Particle Data Group, a difference of just 0.14 per cent. At the time the neutron was discovered in 1932, it was perhaps natural to imagine that it was some kind of composite, consisting of a proton *and* an electron.

Adopting this logic, Heisenberg developed an early theory of the strong force by imagining that this is carried by *electrons*. In this model, protons and neutrons interact and bind together inside the nucleus by exchanging electrons between them, the proton turning into a neutron and the neutron turning into a proton in the process. By the same token, the interaction between two neutrons would involve the exchange of two electrons, one in each 'direction'.

This suggested that the protons and neutrons inside the nucleus are constantly changing identities, flickering back and forth from one to the other. In such a scenario, it makes more sense to think of the proton and neutron as though they are two different 'states' of the same particle, or two sides of the same coin.

Heisenberg introduced a new quantum number to distinguish these states, which he called *isospin*, or 'isotopic spin', in analogy with electron spin. Just like electron spin, he assigned it a fixed value, $I = \frac{1}{2}$. This implies that a nuclear particle with isospin is capable of 'pointing' in two different directions, corresponding to $+\frac{1}{2}$ and $-\frac{1}{2}$. Heisenberg assigned one direction to the proton, the other direction to the neutron. Converting a neutron into a proton would in this picture then be equivalent to 'rotating' its isospin.

It was a crude model, but Heisenberg was able to use it to apply non-relativistic quantum mechanics to the nucleus itself. In a series of papers published in 1932, he accounted for many observations in nuclear physics, such as the relative stability of isotopes and alpha particles (helium nuclei, consisting of two protons and two neutrons). But a model in which the force is carried by electrons is too restricting—it does not allow for any kind of interaction between protons. Experiments soon showed that the strength of the force between protons is comparable to that between protons and neutrons.

Heisenberg's theory didn't survive, but the idea of isospin was seen to hold some promise and was retained. Now, the origin of this quantum number is quite obscure (and despite the name it is not another type of 'spin'—it has nothing to do with the proton or neutron spinning like a top). Today we trace a particle's isospin to the identities of the different kinds of *quarks* from which it is composed (more on this in Chapter 15).

We can now get back to our main story. So, when in 1953 Chinese physicist Chen Ning Yang and American Robert Mills searched for a quantity which is conserved in interactions involving the strong force, they settled on isospin. They then searched for a corresponding symmetry group on which to construct a quantum field theory.

It's clear that the symmetry group U(1) will not fit the bill, as this is limited to one dimension and can describe a field with only one force particle. The interactions between protons and neutrons demand at least three force particles, one positively charged (accounting for the conversion of a neutron into a proton), one negatively charged (accounting for the conversion of a proton into a neutron), and one neutral (accounting for proton–proton and neutron–neutron interactions).

Their reasoning led them to the symmetry group SU(2), the special unitary group of transformations of two complex variables. Again, don't get distracted by this odd label. What's important to note is that the resulting quantum field theory has the right symmetry properties. It introduced a new quantum field analogous to the electromagnetic field in QED. Yang and Mills called it the 'B field'.

Generally speaking, a symmetry group SU(n) has n^2-1 dimensions. In the context of a quantum field theory, the number of dimensions determines the number of force-carrying particles the theory will possess. So, an SU(2) quantum field theory predicts three new force particles (2^2-1), responsible for carrying the force between the protons and neutrons in the nucleus, analogues of the photon in QED. The symmetry group SU(2) thus fits the bill. Yang and Mills referred to the charged force carriers as B^+ and B^-, and the neutral force carrier as B^0 (B-zero). It was found that these carrier particles interact not only with protons and neutrons, but also with each other.

It was here that the problems started. The methods of mass renormalization that had been used so successfully in QED could not be applied to the new theory. Worse still, the zero-interaction term in the perturbation series indicated that the three force particles should all be massless, just like the photon. But this was self-contradictory. The carriers of short-range forces were expected to be heavy particles. Massless force carriers made no sense.

At a seminar delivered at Princeton on 23 February 1954, Yang was confronted by a rather grumpy Pauli. 'What is the mass of this B field?', Pauli wanted to know. Yang didn't have an answer. Pauli persisted. 'We have investigated that question', Yang replied, 'It is a very complex question, and we cannot answer it now.' 'That is not a sufficient excuse', Pauli grumbled.[5]

It was a problem that simply wouldn't go away. Without mass, the force carriers of the Yang–Mills field theory did not fit with physical expectations. If they were supposed to be massless, as the theory predicted, then the strong force would reach well beyond the confines of the nucleus and the force particles should be as ubiquitous as photons, yet they had never been observed. Accepted methods of renormalization wouldn't work. They published a paper describing their results in October 1954. They had made no further progress by this time. Although they understood that the force carriers couldn't be massless, they had no idea where their masses could come from.[6] They turned their attentions elsewhere.

Let's pause for a moment to reflect on this. The language that had developed and which was in common use among physicists in this period is quite telling. Remember that in a quantum field theory the field is the thing, and particles are simply elementary fluctuations or disturbances of the field. Pauli had demanded to know the mass of Yang and Mills' B field. So, an elementary disturbance of a field distributed over space and time is associated with a mass.

Whether we can get our heads around a statement like this is actually neither here nor there as far as the physics is concerned. For as long as quantum field theories of various kinds remain valid descriptions of nature, then this is the kind of description we have to learn to live with. In a quantum field theory the terms in the equation that are associated with mass (called—rather obviously—'mass terms') vary with the *square* of the field and contain a coefficient which is also squared. If this coefficient is identified with the particle mass, m, then the mass terms are related to $m^2\phi^2$, where ϕ (Greek, phi) represents the quantum field in question.[7] Pauli nagged Yang for the mass of the B field because he knew there were *no* mass terms in Yang's equations. He knew that Yang didn't have a satisfactory answer.

And this was only part of the problem. A particle with a spin quantum number s generally has $2s + 1$ ways of aligning (or 'pointing') in a magnetic field, corresponding to $2s + 1$ different values of the magnetic spin quantum number, m_s. Recall that an electron is a fermion with $s = \frac{1}{2}$, and so it has $2 \times \frac{1}{2} + 1 = 2$ different values of m_s ($+\frac{1}{2}$ and $-\frac{1}{2}$) corresponding to spin-up and spin-down states.

What about the photon? Photons are bosons with a spin quantum number $s = 1$. So there are $2 \times 1 + 1 = 3$ possible orientations of the photon spin (three values of m_s), right?

To answer this question, let's play a little game of 'pretend'. Let's forget everything we know about photons and pretend that they're actually tiny, spherical particles or atoms of radiation energy, much as Newton had once envisaged. Let's also forget that they travel only at the speed of light and pretend that we can accelerate them from rest up to this speed. What would we expect to see?

We know from Einstein's special theory of relativity that as the speed of an object increases, from the perspective of a stationary observer time dilates and lengths shorten. As we accelerate a spherical photon, we would expect to see the diameter of the sphere moving in the direction of travel contracting and the particle becoming more and more squashed or oblate. The closer we get to the speed of light the flatter the spheroid becomes. What happens when it reaches light-speed?

Recall from Chapter 5 that the relativistic length l is given by l_0/γ, where l_0 is the 'proper length' (in this case, the diameter of the particle when it is at rest) and γ is $1/\sqrt{(1 - v^2/c^2)}$. When the velocity, v, of the particle reaches c, then γ becomes infinite and l becomes zero. From the perspective of a stationary observer, the particle flattens to a pancake. One with absolutely no thickness.

Of course, photons only travel at the speed of light and our game of pretend is just that—it's not physically realistic. But we are nevertheless able to deduce from this that photons are in some odd way 'two-dimensional'. They are *flat*. Whatever they are they have no dimension—they're *forbidden* by special relativity from having any dimension—in the direction in which they're moving.

Now we can go back to our question about photon spin. If $s = 1$, we would indeed be tempted to conclude that the photon spin can point in three different directions. But, as we've just seen, the photon spin *doesn't have three directions to point in*. By virtue of the simple fact that it is obliged to move at the speed of light, it has only two dimensions. One of the spin directions (one of the possible values of m_s) is 'forbidden' by special relativity.

There are therefore only two possible spin orientations for the photon, not three. These correspond to the two known types of circular polarization, left-circular ($m_s = +1$, by convention), and right-circular ($m_s = -1$). These properties of light may be unfamiliar, but don't fret. We can combine the left-circular and right-circular polarization states of light in a certain kind of superposition which yields linearly polarized states—vertical and horizontal. These are much more familiar (though note that there are still only two—there is no 'back-and-forth' polarization). Polaroid® sunglasses reduce glare by filtering out horizontally polarized light.

So, all massless bosons (such as photons) travel at the speed of light and are 'flat', meaning that they can have only two spin orientations rather than three expected for bosons with $s = 1$. But if, as Yukawa had suggested, the carriers of short-range forces are massive particles then there is no such restriction. Heavy particles cannot travel at the speed of light so they are expected to be 'three-dimensional'. If they are also bosons with $s = 1$, then they would have three spin orientations.

Physicists needed to find a mechanism which would somehow act to slow down the massless force carriers of the Yang–Mills field theory, thereby allowing them to gain 'depth', acquiring a third dimension in the direction of travel for the particles' spin to point in. The mechanism also needed to conjure up mass terms in the equations related to $m^2\phi^2$. Just how *was* this supposed to happen?

Five things we learned

1. As physicists sought quantum field theories to describe the strong and weak forces, they used Noether's theorem: conservation laws are connected with certain continuous symmetries in nature.

2. Heisenberg imagined that the proton and neutron are two different quantum states of a single particle with an *isospin*, $I = \frac{1}{2}$. Interactions in which protons are converted into neutrons and vice versa would then simply involve a 'rotation' of the isospin orientation; for example, from $+\frac{1}{2}$ to $-\frac{1}{2}$. Heisenberg's field theory didn't survive, but the idea of isospin was retained.

3. In seeking to develop a quantum field theory of the strong force, Yang and Mills identified isospin as the conserved quantity and fixed on the symmetry group SU(2), requiring three hypothetical force carriers, which they named as B^+, B^-, and B^0.

4. But their theory predicted that these force carriers should be massless and moving at the speed of light, like photons, completely at odds both with experiment and the expectations for short-range forces. Pauli was not impressed.

5. Physicists needed to find a mechanism which would somehow slow down the hypothetical force carriers, allowing them to gain a third dimension (a third spin orientation) and to conjure up 'mass terms' related to $m^2\phi^2$.

14

THE GODDAMN
PARTICLE

...the publisher wouldn't let us call it the Goddamn Particle,
though that might be a more appropriate title, given its
villainous nature and the expense it is causing. And two, there
is a connection, of sorts, to another book, a much older one...

Leon Lederman[1]

Yang and Mills were in pursuit of a quantum field theory of
the strong force, but what of the weak force? Like the strong
force, the weak force is also short-range, again implying that its
force carriers must be massive particles.

In 1941, Schwinger reasoned that if the weak force is imagined
to be carried by a single force particle,* and this is assumed to be
massive, equal in size to a couple of hundred times the mass of a
proton, then its range would indeed become very limited.[2] Unlike
massless photons, massive particles are very sluggish, moving at
speeds substantially slower than light. A force carried by such
sluggish particles would also be considerably weaker than electromagnetism.

Schwinger realized that if the mass of such a weak force carrier
could be somehow 'switched off'—if the carrier was imagined to

---

* We know now that there are actually three weak force carriers—see later.

be massless—then the weak force would actually have a range and strength similar in magnitude to electromagnetism. This was nothing more than a bit of numerology, and quite an outrageous speculation, but it was also the first hint that it might be possible to *unify* the weak and electromagnetic forces into a single, 'electro-weak' force.[3]

The logic runs something like this. Despite the fact that they appear so very different, the electromagnetic and weak forces are in some strange way manifestations of the same, 'electro-weak', force. They appear very different because something has happened to the carriers of what we now recognize as the weak force. Unlike the photon, these force particles have somehow become 'three-dimensional', their speeds reduced below light-speed, and they have gained a lot of mass. This restricts the range of the force and greatly diminishes its strength relative to electromagnetism.*

We can look at this another way. If we were able to wind the clock back to the very earliest moments of the big bang origin of the universe, then at the energies and temperatures that prevailed at this time all the forces of nature (including gravity, as a 'force' exerted by mass-energy on spacetime) are thought to have been fused together and indistinguishable. Gravity was the first to separate, followed by the strong force. At around a trillionth of a second after the big bang, the 'electro-weak' force split into two separate forces—the weak force and electromagnetism. By this time, all the four forces of nature were established. The key question is this. What happened to the carriers of the weak force to make them so heavy? Or, equivalently, what happened a trillionth of a second after the big bang to cause the electro-weak force to split apart?

---

* At the femtometre length scale of protons and neutrons, the weak force is about 10 million times weaker than electromagnetism.

The challenge was taken up by Schwinger's Harvard graduate student, Sheldon Glashow. After a few false starts, Glashow developed a quantum field theory of weak interactions based, like the Yang–Mills theory, on the SU(2) symmetry group. In this theory the weak force is carried by three particles (remember the number of force carriers involved in an SU(n) structure is determined by $n^2 - 1$). Two of these particles carry electrical charge and are now called the W^+ and W^- particles. This leaves a third, neutral force carrier called the Z^0.

But Glashow now ran into the very same problem that Yang and Mills had experienced. The quantum field theory said that the W and Z particles should all be massless, just like the photon. And if he tried to fudge the equations by adding masses 'by hand', the theory couldn't be renormalized.

So, we know that carriers of the weak force must be massive particles. But the theory says they should be massless. Precisely *how* then do the carriers of the weak force gain their mass? The solution to this puzzle emerged in the seven-year period spanning 1964 to 1971. The answer was to invoke something called *spontaneous symmetry-breaking*.

This sounds all rather grand, but spontaneous symmetry-breaking is actually a familiar, everyday kind of phenomenon. Imagine watching a time-lapse video of a large glass jar of water that is being cooled below freezing. What would we expect to see? At some point we would see the first ice crystals form and then slowly spread through the whole volume of water, turning it eventually into a block of ice.

Now, the water molecules in the liquid have a certain symmetry—they look broadly similar in different directions—up, down, left, right, forward, back—by virtue of their random motions within the loose network that forms the liquid, as shown in Figure 19(a). But ice is a crystal lattice, a regular array with planes tiled (or 'tessellated') with hexagonal rings of atoms. This structure looks distinctly different in different directions. Look to the left or right and we see a 'corridor' formed by the lattice structure. But look up and we see a 'ceiling'; look down and we see a 'floor', as shown in Figure 19(b).

Although the crystal is a more regular, repeating structure, in three dimensions the water molecules are organized in a less symmetrical way than in liquid water—they take up different

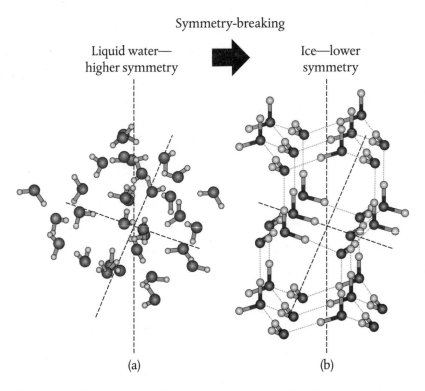

Figure 19. The structure of liquid water is governed by short-range forces between neighbouring molecules, but does not look different in different directions, as we can see in (a). However, ice is a crystal structure, with distinct orientations of the water molecules organized in hexagonal arrays, (b). The ice therefore has a lower overall symmetry compared with the liquid.

orientations in different directions. So freezing the water 'breaks' the higher symmetry of the liquid.

This tells us what spontaneous symmetry-breaking is but not how it works. So let's rewind the video and watch it again, more closely. We see that the first ice crystals form at a specific—but seemingly quite random—place in the volume of water, most likely adjacent to the wall of the jar. Why? We can see that once the first ice crystals have formed, more crystals 'nucleate' around these and the ice expands to fill the jar. So we ask a slightly different question: what causes the first ice crystals to nucleate?

Here's a clue. Let's repeat this experiment with ultra-pure water and ensure that the walls of the jar are perfectly smooth. We now cool the jar really slowly. We find that we can reduce the temperature of the water well below its freezing point without forming any ice at all. This is *super-cooling*. And there's the answer. The first ice crystals nucleate around *impurities* in the water or on *inhomogeneities* in the surface of the jar. So removing the impurities and the inhomogeneities prevents the first ice crystals from forming.

We conclude that the crystals need something to 'hang on to' to get them going. To get spontaneous symmetry-breaking we need to add something (impurities or inhomogeneities, in this case) to encourage it to happen.

How does this help? Well, the SU(2) quantum field theories developed by Yang and Mills and by Glashow are much like the example in which the water is ultra-pure and the walls of the jar perfectly smooth. To get the symmetry to break, physicists realized that they needed to add something—a missing ingredient—to the 'background environment' of the quantum field.

In a sense, they needed something for the massless force carriers of the field to 'hang on to'. This ingredient is needed to break the symmetry and drive a distinction between the forces. Now there really aren't that many options to choose from, so they introduced another, but altogether new, kind of quantum field.

This idea was developed in the early 1960s in connection with the properties of superconducting materials. Japan-born American physicist Yoichiro Nambu realized that spontaneous symmetry-breaking can result in the *formation of particles with mass.*[4]

It took a few years to work out a detailed mechanism. There were hints in the work of Nambu, British theorist Jeffrey Goldstone, and remarks by American physicist Philip Anderson. The mechanism was finally detailed in a series of papers which appeared in 1964, published independently by American physicist Robert Brout and Belgian François Englert, English physicist Peter Higgs at Edinburgh University, and Americans Gerald Guralnik and Carl Hagen and British physicist Tom Kibble, all at Imperial College in London. From about 1972, the mechanism has been commonly referred to as the *Higgs mechanism* and the new quantum field is referred to as the *Higgs field.*

Once again, it is important to remember that theorists are primarily concerned with getting the structure of the mathematics right. They're not overly concerned with the physical interpretation (and certainly not the visualization) of what their mathematical equations are telling them. That's a problem they'll happily leave to somebody else. It was enough that adding a background Higgs field with certain properties did indeed introduce new terms into the equations of the quantum field theory, terms which can be interpreted as mass terms, related to $m^2\phi^2$. The mechanism works in a mathematical sense, and it's left to us to try to make sense of it physically. We can but try.

Adding a background Higgs field suggests that, whatever it is, it pervades the entire universe like a modern-day ether (though much, much more tenuous than the ether of nineteenth-century

physicists such as Maxwell). In the absence of this field, all particles (as it turns out, both matter and force particles) are by default
massless and two-dimensional, and will happily go about their
business at the speed of light.

Make no mistake, if this situation were to persist there would
be no mass and no material substance. There would be no universe of the kind so familiar to us, no stars or galaxies, planets,
life, or *Homo sapiens*. What happens is that the massless particles
interact with the Higgs field, resulting in a number of effects. They
gain a third dimension—they swell and become 'thick'—and
they slow down. As a consequence the particles gain mass
(terms of the form $m^2\phi^2$ appear in the equations)—see Figure 20.
Various analogies have been used to 'explain' these effects, the
most popular suggesting that the Higgs field behaves rather like
molasses, dragging on the particles and slowing them down,
their resistance to acceleration manifesting itself as inertial mass.
Such analogies are always inadequate (Higgs himself prefers to
think of the mechanism as one involving a kind of *diffusion*), but
at least they help us to get our heads around what's happening.

The most important point to remember concerns the 'origin'
of mass. Ever since the Greek atomists, we have tended to think of
mass as an innate, inseparable, 'primary' property of the ultimate
constituents of matter. Galileo and Newton refined this concept
but they did not change it in any essential way. The inertial mass
of an object is a measure of its resistance to acceleration, and our
instinct is to equate inertial mass with the amount of substance
that the object possesses. The more 'stuff' it contains, the harder
it is to accelerate.

We now interpret the extent to which the motion of an otherwise massless elementary particle is 'resisted' by interaction with
the Higgs field as the particle's inertial mass. The concept of mass
has vanished in a puff of mathematical logic. It has become a

(a)

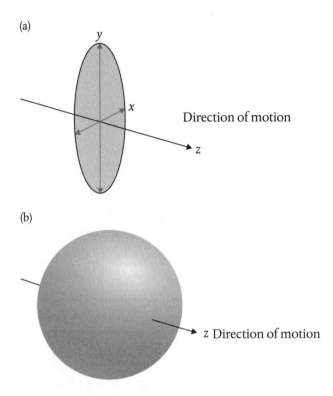

Direction of motion

(b)

z Direction of motion

Figure 20. (a) A massless boson moves at the speed of light and has just two directions to 'point' in, which are shown here as left/right (x) and up/down (y). It cannot 'point' in the direction in which it is moving. (b) On interacting with the Higgs field, the particle gains a third dimension—forward/back (z). The particle gains 'depth', slows down, and mass terms appear in the field equations of the form $m^2\phi^2$.

secondary property, the result of *interactions* between massless particles and the Higgs field.

Looking back, we would now suggest that, a trillionth of a second after the big bang, the temperature of the universe had cooled sufficiently to allow the Higgs field to settle to a fixed value. This provided the background necessary to break the symmetry of the electro-weak force. The W and Z particles found something to 'hang on to'; they gained a third dimension, they gained mass, and the weak force separated from electromagnetism.

Despite its attractions, the Higgs mechanism did not win converts immediately. Higgs actually had some difficulties getting his paper published. He sent it to the European journal *Physics Letters* in July 1964, but it was rejected by the editor as unsuitable. He was indignant, but the simple truth is that in the early 1960s quantum field theory had become rather unfashionable, principally because of the problems that Higgs was now showing how to fix.[5]

Higgs made some amendments to his paper and re-submitted it to the journal *Physical Review Letters*. It was sent to Nambu for peer review. Nambu asked Higgs to comment on the relationship between his paper and a similar article just published in the same journal by Brout and Englert. Higgs had not been aware of Brout and Englert's work on the same problem and acknowledged their paper in an added footnote. He also added a final paragraph to the main text in which he drew attention to the possibility of another, massive boson, the quantum particle of the Higgs field. This would come to be known as the *Higgs boson*.

Physicists now had a mechanism, but not a fully fledged quantum field theory (and certainly not a field theory that could be renormalized). The next step was taken three years later. Steven Weinberg had spent a couple of years puzzling on the effects of spontaneous symmetry-breaking in strong-force interactions when he realized that his approach wasn't going to work. It was at this moment that he was struck by another idea.[6]

Weinberg had been applying the Higgs mechanism to the strong force in an attempt to give mass to the strong force carriers. He now realized that the mathematical structures he had been trying to apply to strong-force interactions were precisely what were needed to resolve the problems with weak-force interactions and the massive force carriers these interactions implied. He had been applying the right idea to the wrong problem—this was the answer to the mystery of massive force carriers in weak-force interactions.[7]

But rather than apply the approach to protons and neutrons, which are also affected by the strong force, Weinberg decided to restrict himself only to particles such as electrons and neutrinos, which are not. Weinberg later confessed his reason. A hesitant Murray Gell-Mann and George Zweig had, a few years earlier, proposed that protons and neutrons are actually composite particles, composed of what would come to be known as quarks (more on this in Chapter 15). Applying the Higgs mechanism to the weak force operating on protons and neutrons would mean involving quarks in the picture, and Weinberg wasn't at all sure these really did exist.[8]

Weinberg published a paper detailing a unified electro-weak theory in November 1967. In this theory the Higgs mechanism works like this. Prior to breaking the symmetry, the electro-weak force is carried by *four* massless particles which, for the sake of simplicity, we will call the W^+, W^0, W^-, and the B^0 particles. Interactions with the background Higgs field cause the W^+ and W^- particles to acquire a third dimension, slow down, and gain mass.

The W^0 and B^0 particles also gain mass, but it's a simple fact in quantum mechanics that neutral particles have a tendency to form superpositions and mix together. The W^0 and B^0 particles mix to produce a massive Z^0 particle and the massless photon. We associate the massive W^+, W^-, and Z^0 particles with the weak force and the massless photon with electromagnetism.

Weinberg was able to estimate the mass-scales of the weak-force carriers. He predicted that the masses of the W particles should be about eighty-five times the proton mass (about 80 billion electron volts, or 80 GeV/c^2) and that the Z^0 should have a mass about ninety-six times that of the proton (about 90 GeV/c^2).

In 1964, Higgs had referred to the possibility of the existence of a Higgs boson, but this was not in relation to any specific force.

In his electro-weak theory, Weinberg had found it necessary to introduce a Higgs field with four components, implying four fundamental field particles (four Higgs bosons). As a result of the interaction, three Higgs bosons are 'swallowed' by the W^+, W^-, and Z^0 particles, adding a third dimension to each and slowing them down.

The fourth appears as a physical particle—a residual Higgs boson.

In the United Kingdom, Abdus Salam had been introduced to the Higgs mechanism by Tom Kibble. He had worked earlier on an electro-weak field theory and immediately saw the possibilities afforded by spontaneous symmetry-breaking. When he saw a preprint of Weinberg's paper he discovered that both he and Weinberg had independently arrived at precisely the same model. He decided against publishing his own work until he had had an opportunity properly to incorporate protons and neutrons in the picture. Both Weinberg and Salam believed that the electro-weak theory was renormalizable, but neither was able to prove this at the time.

The proof followed a few years later. By sheer coincidence, in 1971 Dutch theorists Martinus Veltman and Gerard 't Hooft independently re-discovered the field theory that Weinberg had first developed, but they were now able to show how it could be renormalized. 't Hooft had initially thought to apply the theory to the strong force, but when Veltman asked a colleague about other possible applications, he was pointed in the direction of Weinberg's 1967 paper. Veltman and 't Hooft realized that they had developed a fully renormalizable quantum field theory of electro-weak interactions.

What does this mean for the electron? Recall from Chapter 12 that mass renormalization implies that the mass of the electron has two parts. It has a hypothetical 'bare mass', or the mass it would have if it could be separated from its own, self-generated electromagnetic field. It also has an 'electromagnetic mass', generated by the energy of the interactions between the electron and its electromagnetic field, which 'dress' the electron in a covering of virtual photons.

Now we learn that even the 'bare mass' is not an intrinsic property of the electron. It is derived from interactions between the electron and the Higgs field. These interactions add a third dimension and slow the electron down, resulting in effects that we interpret as mass.

Experimental high-energy particle physics caught up with the theorists a few years later. Weinberg had predicted the masses of the weak-force carriers. At the time he made these predictions there was no particle collider large enough to observe them. But in the years that followed a new generation of particle colliders was constructed in America and at CERN, near Geneva in Switzerland. The discovery of the W particles at CERN was announced in January 1983, with masses eighty-five times that of the proton, just as Weinberg had predicted. The discovery of the Z^0 was announced in June that year, with a mass about 101 times the mass of a proton. (The Z^0 mass is now reported to be ninety-seven times the proton mass.)*

And, of course, the electro-weak theory predicts the existence of the Higgs boson. Given that the Higgs mechanism allows the masses of the weak-force carriers to be predicted with such confidence, the existence of a Higgs field—or something very like it—seemed a 'sure thing'. However, there were alternative theories of symmetry-breaking that did not require a Higgs field and

* So, depending on how you want to interpret the meaning of 'couple', we might conclude that Schwinger's outrageous suggestion was out by a factor of a little more than two.

there remained problems with the electro-weak theory which could not be easily resolved. These problems tended to sow seeds of doubt and erode the theorists' confidence. The Higgs mechanism was far from being proved.

The Higgs mechanism slotted comfortably into the electro-weak theory and rendered it renormalizable. It all seemed to fit together perfectly. But the mechanism demands the existence of a new kind of quantum field which fills all of space. So, it came down to this. If the Higgs field really exists, then so should its fundamental field particle, the Higgs boson.

The obvious next step was to find evidence for the existence of the Higgs boson, and a race began between Fermilab in Chicago and CERN in Geneva. In a book published in 1993, American particle physicist Leon Lederman emphasized (or over-emphasized, depending on your point of view) the fundamental role played by the Higgs boson. He called it the 'God particle'.

He gave two reasons for this name: 'One, the publisher wouldn't let us call it the Goddamn Particle, though that might be a more appropriate title, given its villainous nature and the expense it is causing. And two, there is a connection, of sorts, to another book, a much older one...'.[9]

Five things we learned

1. In 1941, American theorist Julian Schwinger realized that if the carriers of the weak force were imagined to be massless, then the strength and range of the force would be similar to electromagnetism. This was the first clue that these two forces of nature could be *unified*, the weak force and electromagnetism combining to give an 'electro-weak' force.

2. But the weak force and electromagnetism are no longer unified—they are very distinct forces of nature with different strengths and ranges. This means that something must have happened to the carriers of the weak force that led them to gain mass. The force carriers became heavy, considerably reducing the strength and range of the weak force.

3. A mechanism that could do this was developed independently by a number of theorists in 1964, and is today known as the Higgs mechanism. The massless carriers of the weak force interact with the Higgs field, slowing down, gaining a third dimension or spin orientation and gaining mass in the process. In 1967, Weinberg estimated that the weak-force carriers should have masses about eighty-five and ninety-six times the proton mass.

4. The W and Z particles, the carriers of the weak force, were discovered at CERN in 1983. The online database managed by the Particle Data Group gives the masses of the W particles as 80.385 GeV/c^2 (85.7 times the proton mass) and of the Z^0 particle as 91.188 GeV/c^2 (97.2 times the proton mass).

5. But there were other theories that could potentially explain the masses of the W and Z particles which didn't require a new kind of quantum field. To prove that the Higgs field really exists, it would be necessary to find evidence for its fundamental field quantum, the Higgs boson. The race to find the Higgs began.

15

THE STANDARD
MODEL

...the finder of a new elementary particle used to be rewarded
by a Nobel Prize, but such a discovery now ought to be punished
by a $10,000 fine.

Willis Lamb[1]

The decade spanning the late-1960s through to the late 1970s
was a golden age for high-energy particle physics. As the
details of a unified quantum field theory of the weak force and
electromagnetism were being figured out, so too was the nature
of the strong force. By 1979, all the theoretical ingredients of what
was to become known as the *standard model* of particle physics
were in place. This model accounts for the set of all the known
elementary particles and the forces at work inside both the atom
and the atomic nucleus. All that was missing was experimental
confirmation of a few of the heavier members of the set.

The secrets of the strong force were gradually teased out into
the open. Clues were available from studying the pattern of the
'zoo' of particles that experimental physics turned up in the
thirty-year period from the 1930s to the 1960s. To the physicists
of the time, this procession of new particle discoveries was simply bewildering. To the proton, neutron, electron, and neutrino
was added a host of new and bizarrely unusual particles. Dirac's
'dream of philosophers' turned into a nightmare.

In 1932, American physicist Carl Anderson had discovered Dirac's positron in cosmic rays, streams of high-energy particles from outer space that constantly wash over the Earth's upper atmosphere. Four years later, he and fellow American Seth Neddermeyer identified another new particle. It was initially believed that this was one of the particles predicted by Yukawa in 1935—its mass was in the right ballpark. But it behaved just like an electron, with a mass about 200 times greater than an ordinary electron. This particle has had many names and is today called the *muon*. At the time of its discovery it simply didn't fit with any theories or preconceptions of how the building blocks of matter should be organized.

In 1947, another new particle was discovered in cosmic rays by Bristol University physicist Cecil Powell and his team. This was found to have a slightly larger mass than the muon, 273 times that of the electron. But unlike the muon it came in positive, negative and, subsequently, neutral varieties. These were called pi-mesons (or *pions*). *These* were the particles that Yukawa had predicted.

As detection techniques became more sophisticated, the floodgates opened. The pion was quickly joined by positive and negative K-mesons (or *kaons*) and a neutral particle called the *lambda*. New names proliferated. Responding to a question from one young physicist, Fermi remarked: 'Young man, if I could remember the names of these particles, I would have been a botanist.'[2]

The kaons and the lambda in particular behaved very strangely and fit none of the established quantum rules. American physicist Murray Gell-Mann felt he had no alternative but to propose that these particles are governed by some new, hitherto unknown quantum property, which he called *strangeness*,* paraphrasing

---

* Much the same idea was put forward at around the same time by Japanese physicists Kazuhiko Nishijima and Tadao Nakano, who referred to strangeness as 'η-charge'.

Francis Bacon: 'There is no excellent beauty that hath not some strangeness in the proportion.'[3] Whatever this was supposed to be, in strong-force interactions involving particles exhibiting this strange behaviour, their 'strangeness' is conserved.

There had to be some kind of underlying pattern. A structure was needed that would bring order to the zoo and explain how all these weird and wonderful particles were related to each other, much as Russian chemist Dmitry Mendeleev had brought order to the array of chemical elements by arranging them in a periodic table.

Physicists had by this time organized the particles into categories based on their broad properties. There are two principal classes, called *hadrons* (from the Greek *hadros*, meaning 'thick' or 'heavy') and *leptons* (from the Greek *leptos*, meaning small). Hadrons are affected by the strong force, the weak force, and electromagnetism. Leptons are affected only by the weak force and electromagnetism.

The class of hadrons includes a sub-class of *baryons* (from the Greek *barys*, also meaning 'heavy'). These are heavier particles and include the proton, neutron, lambda, and two further series of particles that had been discovered in the 1950s named sigma and xi. The second sub-class is that of *mesons* (from the Greek *mésos*, meaning 'middle'). These particles experience the strong force but are of intermediate mass. Examples include the pions and the kaons. The class of leptons includes the electron, muon, and the neutrino. Both the baryons and the leptons are *fermions*, with half-integral spin quantum numbers.

In the early 1960s, Gell-Mann and Israeli Yuval Ne'eman suggested that the only way to make sense of the pattern of hadrons is to assume that they are not, in fact, elementary. This was history repeating itself. A few centuries of scientific endeavour had led to (admittedly indirect) evidence that matter is composed of atoms. But these were not the ultimate, indivisible atoms of Greek philosophy, or of mechanical philosophy. Atoms are divisible. They contain atomic nuclei and electrons, and the nuclei contain protons and neutrons. Now it was suggested that the protons and neutrons are themselves composites, made up of other particles that are even more elementary.

It slowly dawned on the physicists that these new elementary particles could form the basis for a new understanding of the nature and composition of matter. Together with the leptons they would be the new 'atoms'. They would be the new building blocks from which all the creatures in the particle 'zoo'—protons, neutrons, pions, kaons, lambda, sigma, xi—are assembled.

The pattern suggested that two new kinds of elementary particle would be needed to make up a proton or a neutron, with each containing *three* of these. For example, if we called the two different kinds of new particles A and B, then the proton and neutron appeared to require combinations such as AAB and ABB. But these are quite awkward combinations. What do we do about the proton's electric charge? Do we suppose that one of these new particles carries the positive charge (B^+, say), with the other particle neutral? But then simple combinations of these cannot reproduce the electrically neutral neutron. The combination AAB^+ gives us a positive proton, but AB^+B^+ doesn't give us a neutral neutron. This kind of pattern doesn't fit.

This problem was highlighted by Gell-Mann himself, when his colleague Robert Serber suggested precisely this three-of-two-kinds solution over lunch at Columbia University in New York in 1963. Gell-Mann was dismissive. 'It was a crazy idea', he said. 'I grabbed the back of a napkin and did the necessary calculations to show that to do this would mean that the particles would have to have fractional electric charges— $-\frac{1}{3}$, $+\frac{2}{3}$, like so—in order to add up to a proton or neutron with a charge of plus [one] or zero.'[4]

This is the only way the electric charge can be distributed over three particles of two kinds. To make a proton we need two particles with a charge of $+\frac{2}{3}$ and one particle carrying $-\frac{1}{3}$ to get an overall charge of $+1$. One particle with a charge of $+\frac{2}{3}$ and two carrying $-\frac{1}{3}$ would give a neutron with overall zero charge. To account for other particles in the zoo a third kind of particle, also with a charge of $-\frac{1}{3}$, was needed.

There were certainly no precedents for thinking that particles could possess fractional electric charges, and the whole idea seemed ridiculous. Gell-Mann called them 'quorks', a nonsense word deliberately chosen to highlight their absurdity. But despite these worrying implications, there was no doubting that this combination did provide a potentially powerful explanation for the pattern. Maybe if the 'quorks' are forever trapped or *confined* inside the larger hadrons, then this might explain why fractionally charged particles had never been seen in high-energy physics experiments.

As he wrestled with this idea, Gell-Mann happened on a passage from James Joyce's *Finnegan's Wake*: 'Three quarks for Muster Mark!'[5] The word quark didn't quite rhyme with his original 'quork' but it was close enough. Gell-Mann now had a name for these odd new particles.*

The pattern demanded three different kinds of quarks, which Gell-Mann called 'up' (u), with a charge of $+\frac{2}{3}$, 'down' (d), with a charge of $-\frac{1}{3}$, and strange (s), a heavier version of the down quark also with a charge of $-\frac{1}{3}$. The baryons known at that time could then be formed from various permutations of these three quarks and the mesons from combinations of quarks and anti-quarks.

---

* At around the same time, American physicist George Zweig developed an entirely equivalent scheme based on a fundamental triplet of particles that he called 'aces'. Zweig struggled to get his papers published, but Gell-Mann subsequently made strenuous efforts to ensure Zweig's contributions were recognized.

The properties up, down, and strange are described as types of quark 'flavour', which we can think of as a new kind of quantum number.* This is obviously not intended to suggest that quarks actually possess flavour as we experience it. It's best to think of quark flavour as a property similar to electrical charge. Quarks and leptons possess electrical charge and this comes in two varieties—positive and negative. In addition to electrical charge, quarks also possess flavour, and at the time of these proposals there were three varieties—up, down, and strange. We now know that there are actually six flavours.

In this scheme the proton consists of two up quarks and a down quark (uud), with a total electrical charge that adds up to +1. The neutron consists of an up quark and two down quarks (udd), with a total charge that balances out at zero. Beta radioactivity could now be understood to involve the conversion of a down quark in a neutron into an up quark, turning the neutron into a proton, with the emission of a W^- particle.

Incidentally, recall from Chapter 13 that Heisenberg had tried to develop an early quantum field theory based on the idea that the proton and neutron are two different 'states' of the same particle. We can now see that this idea contained more than a grain of truth. Protons and neutrons are indeed composed of the same two flavours of quark—up and down—and differ by virtue of the fact that in the neutron an up quark has been replaced by a down quark.

Isospin is now defined as half the number of up quarks minus the number of down quarks.† For the neutron, this gives an

---

* This doesn't appear to make a lot of sense, as the flavours are labels ('up', 'down',...) rather than numbers (1, 2,...). We can rescue the situation by supposing that an up quark has an up quantum number of 1, and so on for the other flavours.

† The relation is a little bit more involved than this. In fact, the isospin is given as half × (number of up quarks minus number of anti-up quarks) minus (number of down quarks minus number of anti-down quarks).

isospin of ½ × (1 − 2), or −½. 'Rotating' the isospin of the neutron is then equivalent to changing a down quark into an up quark, giving a proton with an isospin of ½ × (2 − 1), or +½.

The kaons and the lambda behave 'strangely' because they contain strange quarks. The positive kaon is a meson formed from up and anti-strange quarks, the negative kaon from strange and anti-up quarks, and the neutral kaon is a superposition of down-anti-strange and strange-anti-down (remember, electrically neutral particles have a tendency to mix together). The lambda is a baryon formed from a combination of up, down, and strange quarks. It is a kind of 'heavy neutron', in which a down quark in a neutron is replaced with a strange quark.

Hints that there may be a fourth quark emerged in 1970. This is a heavy version of the up quark with charge +⅔. It was called the charm quark. Most physicists were sceptical. But when another new particle, called the J/ψ (pronounced 'jay-psi'), was discovered in the 'November revolution' of 1974, simultaneously at Brookhaven National Laboratory in New York and the Stanford Linear Accelerator Center in California, it was realized that this is a meson formed from charm and anti-charm quarks.* The scepticism vanished.

It was now understood that the neutrino is partnered with the electron (and so it is now called the electron-neutrino). The muon-neutrino was discovered in 1962, and for a time it seemed possible that the elementary building blocks of material substance are formed into two 'generations' of matter particles. The up and down quarks, and the electron and electron-neutrino

---

* This seems like an odd name for a particle, but it reflects the simple fact that it was discovered near-simultaneously by two different laboratories. Physicists at Brookhaven National Laboratory called it the 'J'; physicists at Stanford Linear Accelerator Center called it the 'ψ'. In the subsequent tussle for precedence neither group was prepared to concede, so the particle is today called the J/ψ.

form the first generation. The charm and strange quarks, muon and muon-neutrino form a heavier second generation.

When the discovery of another, even heavier, version of the electron—called the tau—was announced in 1977, this caused less consternation than you might imagine. It was quickly assumed that there must be a *third* generation of matter particles, implying the existence of another pair of heavier quarks and a tau-neutrino. American physicist Leon Lederman found evidence for what came to be known as the bottom quark at Fermilab in Chicago in August 1977. He and his colleagues discovered the *upsilon*, a meson consisting of a bottom quark and its anti-quark. The bottom quark is an even heavier, third-generation version of the down and strange quarks, with a charge of $-\frac{1}{3}$.

The discoveries of the top quark and the tau-neutrino were announced at Fermilab in March 1995 and July 2000, respectively. Together they complete the heavier third generation of matter particles. Although further generations of particles are not impossible, there are some reasonably compelling arguments from theory and some experimental evidence to suggest that three generations is probably all there is.

The quark model was a great idea, but at the time these particles were proposed there was simply no experimental evidence for their existence. Gell-Mann was himself rather cagey about the status of his invention. He had argued that the quarks are somehow 'confined' inside their larger hosts and, wishing to avoid getting bogged down in philosophical debates about the reality or otherwise of particles that could never be seen, he referred to them as 'mathematical'.

But experiments carried out at the Stanford Linear Accelerator Center in 1968 provided strong hints that the proton is indeed a composite particle containing point-like constituents inside it. It was not clear that these constituents were necessarily quarks and the results suggested that, far from being held tightly inside the proton, they actually rattle around as though they are entirely free. This seemed to contradict the idea that the quarks are never seen because they are forever bound or confined inside protons and neutrons. If they're free to rattle around, why don't they ever come out?

The solution to this puzzle is actually breathtakingly simple. Our instinct is to think of a force of nature as something that is centred on a point—typically the centre of a particle or an object which 'generates' the force, and which declines in strength the further away we get from it. The obvious examples are Newtonian gravity and electromagnetism, both of which are forces that decline by $1/r^2$ as the distance r from the centre increases, as shown in Figure 21(a).* Hurl it far enough from the Earth's centre, and a rocket will escape Earth's gravitational force. The pull we feel between the north and south poles of two bar magnets gets weaker as we move them apart.

But, as Princeton theorists David Gross and Frank Wilczek and Harvard theorist David Politzer showed in 1973, the strong force doesn't behave this way at all. Instead, it acts as though adjacent quarks are tethered to each other by a piece of strong elastic or a spring—Figure 21(b). As the quarks move close together, the elastic or the spring slackens, and the force between them diminishes. Inside a proton or neutron, the quarks are tethered but close enough together to be free to 'rattle around'.

But the strong force is like a sleeping tiger. Trying to pull the quarks apart is then like tugging on the tiger's tail. The strong force awakens,† and we feel its full strength as the elastic or the

---

* Although it is not immediately obvious, the $1/r^2$ behaviour is related to the simple fact that space has three dimensions.

† A pun intended for Star Wars fans.

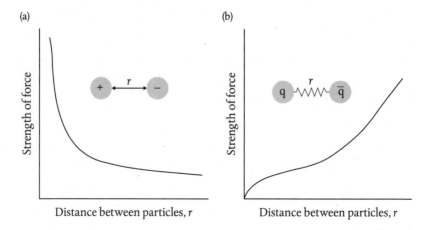

Figure 21. The electromagnetic force between two electrically charged particles increases in strength as the particles move closer together, (a). But the colour force that binds quarks together behaves differently, (b). In the limit of zero separation between a quark and an anti-quark (for example), the force falls to zero. The force increases as the quarks are separated.

spring tightens. It resists further separation and the force grows stronger. The strong force works in the opposite sense to Newtonian gravity and electromagnetism.

Gell-Mann, German theorist Harald Fritzsch, and Swiss theorist Heinrich Leutwyler now had all the ingredients to develop a quantum field theory of the strong force. But this is not a strong force acting between protons and neutrons, which had preoccupied Yang and Mills in the early 1950s. It is instead an even stronger force acting between quarks *inside* protons and neutrons.

The same questions that had confronted Yang and Mills now had to be confronted once again. In the even-stronger strong force interactions between quarks, what property is conserved? And what kind of symmetry is required?

Yang and Mills had fixed on isospin and the SU(2) symmetry group, but their focus had been on interactions between protons and neutrons. Quark flavour was ruled out because the pattern of quarks inside a proton or neutron demand that each hold two quarks of the same flavour (two up quarks in a proton and two down quarks in a neutron). If, as seemed reasonable, the quarks are also spin-½ fermions, then the Pauli exclusion principle applies—a proton or neutron cannot comfortably hold two quarks in the same quantum state.

Yet another new quantum number was needed. Gell-Mann and Fritzsch had earlier fixed on the idea of 'colour'.[6] In this model, each quark possesses one of three different kinds of 'colour charge'—red, green, or blue. Baryons are formed from three quarks each of different colour, such that their total colour charge balances out to zero (or 'neutral') and the resulting particle is 'white'.* For example, a proton may consist of a blue up quark, a red up quark, and a green down quark. A neutron may consist of a blue up quark, a red down quark, and a green down quark. The mesons, such as pions and kaons, consist of coloured quarks and their anti-coloured anti-quarks, such that the total colour charge is zero or neutral and the particles are also 'white'.

So, in these even-stronger strong force interactions, it is quark colour that is conserved. The fact that there are three different colours means that the quantum field theory has to be constructed based on the symmetry group SU(3), the special unitary group of transformations of three complex variables. Eight force carriers are required (remember the number of force particles in an SU(n) field theory is given by n^2-1). These force carriers are called *gluons*, the particles which 'glue' the coloured quarks together inside hadrons.

We can now stop referring to an 'even stronger' strong force and simply call it the *colour force*. Gell-Mann called the resulting field theory *quantum chromodynamics*, or QCD.

---

* Actually, 'colourless' would be more accurate.

But, hang on. Earlier logic demanded that forces operating over very short distances must be carried by massive particles. So does this mean that the gluons are massive? Actually, no. This logic applies to a short-range force like the weak force, which declines in strength as the distance between particles being acted on increases. But, as we just saw, the colour force operates in a very different way. The gluons are massless particles, like the photon, but unlike photons they cannot 'leak' from inside the nucleus of an atom. Like the quarks, they also carry colour charge and are 'confined'.

So what about Yukawa's prediction that the strong force should be carried by particles with a mass of around 100 MeV/c^2? It turns out he was partly right. Although it stays within the nucleus, the colour force binding quarks together inside protons and neutrons strays beyond the confines of these particles, giving rise to a 'residual' strong force. This can then be imagined to act between the protons and neutrons in the nucleus, binding them together. As there are only two particles to be acted on, only three residual force carriers are required. These are, in fact, the pions, with masses of 139.6 MeV/c^2 (π^+ and π^-) and 135.0 MeV/c^2 (π^0).*

So, this is the standard model of particle physics. It consists of QCD and an electro-weak field theory which is split by the Higgs mechanism into an SU(2) field theory of the weak force (which is sometimes—though rarely—referred to as quantum flavourdynamics, QFD) and the U(1) field theory of quantum electrodynamics, or QED. It consists of three generations of matter

---

* It's important to emphasize the word 'residual'. This is *not* a principal force of nature. The pions possess a spin quantum number of 0 so they're not force carriers in the 'proper' sense (they are often referred to as 'Nambu-Goldstone bosons'). Nevertheless, the residual strong force is crucial. It serves to bind protons and neutrons together inside atomic nuclei through the exchange of pions.

particles, a collection of force particles, and the Higgs boson. By 2000, all the particles involved had been discovered experimentally, except for the Higgs.

Fermilab's Tevatron particle collider promised that hints of the Higgs might be glimpsed, but there were no guarantees and it was clear that the Tevatron couldn't create sufficiently high collision energies to yield a convincing discovery. In 1986, American physicists embarked on an ambitious project to build the world's largest particle collider, known as the Superconducting Supercollider. But the project was cancelled by the US Congress in October 1993, with nothing to show for the $2 billion that had already been spent except for a rather large hole beneath the Texas prairie. All hopes transferred to a new collider, commissioned just over a year later, to be built at CERN in Geneva, called the Large Hadron Collider (LHC).

The LHC has two main detectors, called ATLAS and CMS, each of which involves collaborations of about 3,000 physicists from around the world. The first proton–proton collisions were recorded at the LHC on 30 March 2010, at a collision energy of 7 trillion electron volts (7 TeV). This was half the LHC's design energy, but still the highest energy particle collisions ever engineered on Earth. Excitement built through the next two years as collision data at 7 TeV, then 8 TeV, were accumulated. The discovery of the Higgs boson, with a mass of about 125 GeV/c^2 (equivalent to about 133 protons) was announced on 4 July 2012. I watched events unfold in a webcast live from CERN's main auditorium.*

Further data were collected through to mid-December 2012 and at a conference in March 2013 the results from both detector

---

* This is recounted in Jim Baggott, *Higgs: The Invention and Discovery of the 'God Particle'*, Oxford University Press, 2012.

collaborations confirmed the identity of the new particle. This was definitely a Higgs boson, although the physicists remained deliberately vague about it, declaring '…we still have a long way to go to know what kind of Higgs boson it is'.[7] But this was more than enough for the Nobel Prize committee, which awarded the 2013 Physics Prize to Higgs and Englert.*

The physicists' confidence that this is indeed 'the' standard model Higgs boson continued to grow. When the results from both collaborations were combined and evaluated in a paper published in September 2015 it was clear that the new particle is entirely consistent with the predictions of the standard model.[8] The standard model of particle physics was now complete (Figure 22).

In December 2015, the LHC, now operating at a collision energy of 13 TeV, produced tantalizing hints of a new, beyond-standard model particle with a mass of about 750 GeV/c^2. This caused quite a buzz and had theorists scrambling to churn out papers at a rate of about ten per week. But as further experimental data were gathered at the LHC in 2016, these 'hints' were revealed to be nothing more than rather cruel, misleading statistical fluctuations. At the time of writing, there is no observation or experimental result in high-energy particle physics that cannot be accommodated within the standard model framework.

The loss of the 750 GeV/c^2 'hint' caused considerable grief within the particle physics community. This might seem odd, given that the discovery of the Higgs boson proved to be such a triumph. The simple truth is that physicists are desperate for some guidance from experiment on how the standard model might be transcended.

---

* Robert Brout sadly died in 2011 after a long illness, and the Nobel Prize cannot be awarded posthumously.

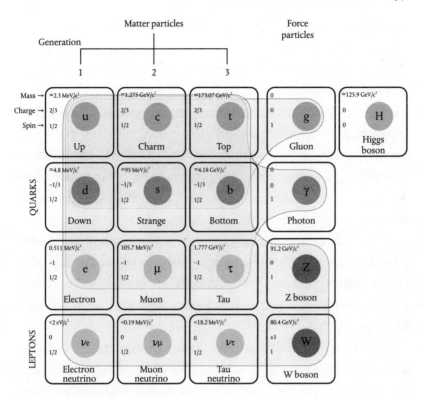

Figure 22. The standard model of particle physics describes the interactions of three generations of matter particles through three kinds of force, mediated by a collection of 'force carriers'. The masses of the matter and force particles are determined by their interactions with the Higgs field.

Why so? Put simply, the standard model is full of glaring omissions. As it stands, the model provides no clues about the strengths of the interactions between the elementary matter and force particles and the Higgs field, and so it cannot be used to calculate the masses of these particles from 'first principles'. The Higgs mechanism tells us that this is where mass comes from, but it cannot tell us how much mass will result. The particle masses (or the strengths of their interaction with the Higgs field) have to be put in 'by hand' based on experimental measurements.

All the particles in the standard model have anti-matter counterparts. Anti-particles possess the same masses as their matter equivalents but have opposite electrical charges, such as the positron and electron (neutral particles are their own anti-particles). When they collide, particles and anti-particles will annihilate to produce high-energy photons (gamma rays). But, provided they are kept separate from matter, the anti-particles are perfectly stable.

Anti-hydrogen atoms, consisting of a negatively charged anti-proton and a positively charged positron, have been produced at CERN and trapped for short periods to allow them to be studied. They have physical properties essentially identical to ordinary hydrogen atoms.

So, why is the visible universe made of matter, and not anti-matter (or a mixture)? If equal amounts of matter and anti-matter were produced in the first few moments following the big bang, as seems reasonable, why did it not all annihilate, leaving a universe full of light but empty of matter? The arm-waving explanation is that—by chance or necessity—as the early universe evolved, the balance was tipped ever so slightly in favour of matter particles. The standard model gives no clue as to why this might have happened.

Then there's the puzzle of dark matter. As we learned in Chapter 8, to explain the large-scale structure of the universe as it appears to us today we need to invoke another form of matter which is detectable by virtue of its gravity but invisible to all forms of electromagnetic radiation. We can't see it but we know it must be there. None of the elementary building blocks of the standard model meet the requirements of dark matter. We have no idea what it is.

And, finally, there is no room in the standard model for the 'force' of gravity, essential for our description of matter on a large scale. Einstein's general theory of relativity works extraordinarily well, as we've seen. But the standard model of particle physics is constructed from a set of quantum field theories. Now, the general theory of relativity handles the large-scale behaviour of mass-energy and curved spacetime. Quantum field theories handle the colour-force, weak-force and electromagnetic interactions of atomic and sub-atomic particles. When we try to put these two theoretical structures together to create some kind of unified theory that could do the work of both, we find that they really don't get along.

There are differing views about why this might be, but it seems clear that general relativity and quantum field theory treat space and time in ways that are quite different, if not contradictory. In general relativity spacetime is *active*; it *results* from interactions involving matter and energy. In quantum field theory spacetime is *passive*; it merely provides a background in which interactions involving matter and force particles take place. Theorists have been trying to find ways to fix this for more than forty years, and we will take a brief look at their efforts in the epilogue, but it is fair to say that progress has been slow and there is no real consensus on the way forward.

There's clearly plenty still to do. But, in the meantime, I think we're now is a position to address the question with which I opened this book. What, exactly, *is* matter?

Five things we learned

1. In the period from the late 1960s to the early 1970s, physicists struggled to discover a principle that would help them to

make sense of the 'zoo' of particles that had been discovered, many exhibiting some strange and exotic behaviour.

2. The idea that hadrons, such as protons and neutrons, might be composite particles consisting of even more elementary quarks was initially dismissed as absurd. Such quarks would have to possess fractional electric charges, with values $+\frac{2}{3}$ and $-\frac{1}{3}$.

3. Experimental evidence that protons and neutrons are indeed composite particles was found in 1968, and when it was discovered that the strong force acts very differently to the weak force and electromagnetism, the idea of quarks began to become more palatable.

4. We now know that there are three generations of matter particles. The first generation consists of up and down quarks (from which protons and neutrons are made), electrons and electron-neutrinos. The particles of all three generations had been discovered by 2000. The discovery of the Higgs boson in 2012 completed the set required by the standard model of particle physics.

5. But there are many things that the standard model cannot explain, such as the strengths of the interactions of matter and force particles with the Higgs field (and hence the particle masses). And there are no real clues as to how the standard model might be transcended.

16

MASS WITHOUT MASS

The origin of the bulk of the mass of ordinary matter is well accounted for, in a theory based on pure concepts and using no mass parameters—indeed, no mass unit—at all!

Frank Wilczek[1]

We're almost there. In this final chapter, I propose to address the principal question concerning the nature of matter in the context of a familiar, everyday substance—water. Well, water in a very specific form. Imagine a cube of ice, measuring a little over one inch (or 2.7 centimetres) in length. Imagine holding this cube of ice in the palm of your hand. It is cold, and a little slippery. It weighs hardly anything at all yet we know it weighs *something*.

We make our question a little more focused. What is this cube of ice made of? And, an important secondary question: What is responsible for its mass?

To the ancient Greek philosophers, water was one of the four elements along with earth, air, and fire. The Greek atomists argued that material substance cannot suddenly appear from nothing and cannot be divided endlessly into nothing. The atoms of the Greeks possessed specific properties, of size, shape, position, and weight (or mass).

Our cube of ice must therefore be composed of atoms, and atoms cannot exist except within empty space, called the void. Lucretius argued that the very fluid properties of water imply that it contains small atoms with a 'readiness to roll', in contrast with honey, which is stickier and must consist of atoms that are not so smooth, or so fine, or so round.[2] But as the environment surrounding the water is cooled it expels particles or atoms of heat, thereby cooling and eventually freezing the water.[3] We might therefore imagine that as water freezes to ice, its small, spherical atoms are drawn closer and closer together, eventually to lock ranks and form the regular array characteristic of a solid.

The atoms of the mechanical philosophers—of Bacon, Boyle, and Newton, among others—were not much more sophisticated. In Querie 31 of his *Opticks*, Newton speculated that *forces* might be at work between the atoms. But he would not be drawn on the precise nature of these forces beyond suggesting that they might be already known to us, in the forms of gravity, electricity, and magnetism.

To understand what a cube of ice is made of, we need to draw on the learning acquired by the chemists. Building on a long tradition established by the alchemists, these scientists distinguished between different chemical elements, such as hydrogen, carbon, and oxygen. Research on the relative weights of these elements and the combining volumes of gases led Dalton and Gay-Lussac to the conclusion that different chemical elements consist of atoms with different weights which combine according to a set of rules involving whole numbers of atoms.

The mystery of the combining volumes of hydrogen and oxygen gas to produce water was resolved when it was realized that hydrogen and oxygen are both diatomic gases, H_2 and O_2. Water is then a compound consisting of two hydrogen atoms and one oxygen atom, H_2O.

This partly answers our first question. Our cube of ice consists of molecules of H_2O organized in a regular array. We can also make a start on our second question. Avogadro's law states that a mole of chemical substance will contain about 6×10^{23} discrete 'particles'. Now, we can interpret a mole of substance simply as its molecular weight scaled up to gram quantities. Hydrogen (in the form of H_2) has a relative molecular weight of 2,* implying that each hydrogen atom has a relative atomic weight of 1. Oxygen (O_2) has a relative molecular weight of 32, implying that each oxygen atom has a relative atomic weight of 16. Water (H_2O) therefore has a relative molecular weight of $2 \times 1 + 16 = 18$.

It so happens that our cube of ice weighs about 18 grams, which means that it represents a mole of water, more or less.[4] According to Avogadro's law it must therefore contain about 6×10^{23} molecules of H_2O. This would appear to provide a definitive answer to our second question. The mass of the cube of ice derives from the mass of the hydrogen and oxygen atoms present in 6×10^{23} molecules of H_2O.

But, of course, we can go further. We learned from Thompson, Rutherford, and Bohr and many other physicists in the early twentieth century that all atoms consist of a heavy, central nucleus surrounded by light, orbiting electrons. We subsequently learned that the central nucleus consists of protons and neutrons. The number of protons in the nucleus determines the chemical identity of the element: a hydrogen atom has one proton, an oxygen atom has eight (this is called the *atomic number*). But the total mass or weight of the nucleus is determined by the total number of protons *and* neutrons in the nucleus.

Hydrogen still has only one (its nucleus consists of a single proton—no neutrons). The most common isotope of oxygen

* For example, relative to the weight of hydrogen in hydrogen chloride, HCl.

has—guess what—sixteen (eight protons and eight neutrons). It's obviously no coincidence that these proton and neutron counts are the same as the relative atomic weights I quoted above.

If we ignore the electrons, then we would be tempted to claim that the mass of the cube of ice resides in all the protons and neutrons in the nuclei of its hydrogen and oxygen atoms. Each molecule of H_2O contributes ten protons and eight neutrons, so if there are 6×10^{23} molecules in the cube and we ignore the small difference in mass between a proton and a neutron, we conclude that the cube contains in total about eighteen times this figure, or 108×10^{23} protons and neutrons.

So far, so good. There is nothing in this calculation that the Greek atomists or mechanical philosophers would dispute. Yes, our understanding of the nature and composition of matter is now a lot more sophisticated, but the conclusions are essentially the same. For 'atoms' read 'elementary particles'. Setting aside the small contribution from all the electrons, we trace the mass of our cube of ice to the mass of all the protons and neutrons it contains, as shown in Figure 23.

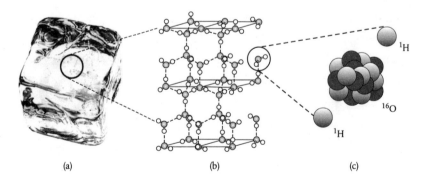

(a)　　　　　　　　(b)　　　　　　　　(c)

Figure 23. A cube of ice measuring 2.7 cm in length will weigh about 18 g, (a). It consists of a lattice structure containing a little over 600 billion trillion molecules of water, H_2O, (b). Each atom of oxygen contains eight protons and eight neutrons, and each hydrogen atom contains one proton, (c). The cube of ice therefore contains about 10,800 billion trillion (108×10^{23}) protons and neutrons.

But we're not quite done yet. We now know that protons and neutrons are not elementary particles. They consist of quarks. A proton contains two up quarks and a down quark, a neutron two down quarks and an up quark. And the colour force binding the quarks together inside these larger particles is carried by massless gluons.

Okay, so surely we just keep going. If once again we approximate the masses of the up and down quarks as the same we just multiply by three and turn 108×10^{23} protons and neutrons into 324×10^{23} up and down quarks. We conclude that *this* is where all the mass resides. Yes?

No. This is where our naïve atomic preconceptions unravel. We can look up the masses of the up and down quarks on the Particle Data Group website. The up and down quarks are so light that their masses can't be measured precisely and only ranges are quoted. The following are all reported in units of MeV/c^2. In these units the mass of the up quark is given as 2.3 with a range from 1.8 to 3.0. The down quark is a little heavier, 4.8, with a range from 4.5 to 5.3.* Compare these with the mass of the electron, about 0.51 measured in the same units.

Now comes the shock. In the same units of MeV/c^2 the proton mass is 938.3, the neutron 939.6. The combination of two up quarks and a down quark gives us only 9.4, or just one per cent of the mass of the proton. The combination of two down quarks and an up quark gives us only 11.9, or just 1.3 per cent of the mass of the neutron. About ninety-nine per cent of the masses of the proton and neutron seem to be unaccounted for. What's gone wrong?

To answer this question, we need to recognize what we're dealing with. Quarks are not self-contained 'particles' of the kind

---

* I should point out that these are very 'messy' estimates, which is why ranges are quoted. But, whilst there is considerable uncertainty, we can be quite confident that the down quark is heavier than the up quark.

that the Greeks or the mechanical philosophers might have imagined. They are quantum wave-particles; fundamental vibrations or fluctuations of elementary quantum fields. The up and down quarks are only a few times heavier than the electron, and we've demonstrated the electron's wave-particle nature in countless laboratory experiments. We need to prepare ourselves for some odd, if not downright bizarre behaviour.

And let's not forget the massless gluons. Or special relativity, and $E = mc^2$. Or the difference between 'bare' and 'dressed' mass. And, last but not least, let's not forget the role of the Higgs field in the 'origin' of the mass of all elementary particles. To try to understand what's going on inside a proton or neutron we need to reach for quantum chromodynamics, the quantum field theory of the colour force between quarks.

Quarks and gluons possess colour 'charge'. Just what is this, exactly? We have no way of really knowing. We do know that colour is a property of quarks and gluons and there are three types, which physicists have chosen to call red, green, and blue. But, just as nobody has ever 'seen' an isolated quark or gluon, so more or less by definition nobody has ever seen a naked colour charge. In fact, quantum chromodynamics (QCD) suggests that if a colour charge could be exposed like this it would have a near-infinite energy. Aristotle's maxim was that 'nature abhors a vacuum'. Today we might say: 'nature abhors a naked colour charge'.

So, what would happen if we could somehow create an isolated quark with a naked colour charge? Its energy would go up through the roof, more than enough to conjure virtual gluons out of 'empty' space. Just as the electron moving through its own self-generated electromagnetic field gathers a covering of virtual photons, so the exposed quark gathers a covering of virtual gluons. Unlike photons, the gluons themselves carry colour charge and they are able to reduce the energy by, in part, masking the exposed colour charge. Think of it this way: the naked quark is acutely embarrassed, and it quickly dresses itself with a covering of gluons.

This isn't enough, however. The energy is high enough to produce not only virtual particles (like a kind of background 'noise' or 'hiss'), but elementary particles, too. In the scramble to cover the exposed colour charge, an anti-quark is produced which pairs with the naked quark to form a meson. A quark is never—but *never*—seen without a chaperone.

But this *still* doesn't do it. To cover the colour charge completely we would need to put the anti-quark in precisely the same place at precisely the same time as the quark. Heisenberg's uncertainty principle won't let nature pin down the quark and anti-quark in this way. Remember that a precise position implies an infinite momentum, and a precise rate of change of energy with time implies an infinite energy. Nature has no choice but to settle for a compromise. It can't cover the colour charge completely but it can mask it with the anti-quark and the virtual gluons. The energy is at least reduced to a manageable level.

This kind of thing also goes on inside the proton and neutron. Within the confines of their host particles, the three quarks rattle around relatively freely. But, once again, their colour charges must be covered, or at least the energy of the exposed charges must be reduced. Each quark produces a blizzard of virtual gluons that pass back and forth between them, together with quark–anti-quark pairs. Physicists sometimes call the three quarks that make up a proton or a neutron 'valence' quarks, as there's enough energy inside these particles for a further 'sea' of quark–anti-quark pairs to form. The valence quarks are not the only quarks inside these particles.

What this means is that the mass of the proton and neutron can be traced largely to the *energy* of the gluons and the sea of quark–anti-quark pairs that are conjured from the colour field.

How do we know? Well, it must be admitted that it is actually really rather difficult to perform calculations using QCD. The colour force is extremely strong, and the corresponding energies of colour-force interactions are therefore very high. Remember that the gluons also carry colour charge, so everything interacts with everything else. Virtually anything can happen, and keeping track of all the possible virtual and elementary-particle permutations is very demanding.

This means that although the equations of QCD can be written down in a relatively straightforward manner, they cannot be solved analytically, 'on paper'. Also, the mathematical sleight-of-hand used so successfully in QED no longer applies—because the energies of the interactions are so high we can't apply the techniques of renormalization. Physicists have had no choice but to solve the equations on a computer instead.

Considerable progress was made with a version of QCD called 'QCD-lite'. This version considered only massless gluons and up and down quarks, and further assumed that the quarks themselves are also massless (so, literally, 'lite'). Calculations based on these approximations yielded a proton mass that was found to be just ten per cent lighter than the measured value.

Let's stop to think about that for a bit. A simplified version of QCD in which we assume that no particles have mass to start with nevertheless predicts a mass for the proton that is ninety per cent right. The conclusion is quite startling. *Most of the mass of the proton comes from the energy of the interactions of its constituent quarks and gluons.*

Since these calculations were performed further progress has been made with a version of the theory called lattice QCD. The magnitudes of the quantum fields representing the quarks are defined only at specific points on a three-dimensional grid or lattice (rather than continuously through spacetime, as would be required for a continuous field). The magnitudes of the gluon fields are then defined on the links connecting neighbouring points on the lattice. The spacing between lattice points is of the order of a few tenths to a few hundredths of a femtometre (10^{-15} metres), and the more this is reduced the closer we get to a 'continuum' version of QCD. To reduce the number of computer calculations required, physicists perform calculations at smaller and smaller spacings and then extrapolate their results to zero spacing.

In lattice QCD it becomes possible to relax the approximations that were required in QCD-lite. We can involve more generations of quarks and include their masses. Great accuracy is possible but this comes at a cost; the most rigorous lattice QCD calculations require the world's largest supercomputers.

If all this isn't bad enough, we must remember that quarks also carry electrical charge and, just like the electron, there will be contributions to the mass arising from the quarks' interactions with their own self-generated electromagnetic fields. This can be treated using QED and the renormalization methods pioneered in the late 1940s.

Lattice QCD calculations reported in November 2008 yielded a proton mass that is within a few per cent of the measured value.[5] Lattice QCD and combined QED calculations of the small mass difference between the proton and the neutron were reported in March 2015.[6] The latter results are particularly interesting. If we assume that the electrical charge carried by the proton is uniformly distributed, then we might be tempted to conclude that, just like the electron, interactions with its own electromagnetic field would add an 'electromagnetic mass' to the proton that the neutron simply can't possess. In the absence of any other effects, we would conclude that the proton must therefore be a little heavier than the neutron. But the opposite is true—the neutron is in fact a little heavier than the proton.

Let's dig a little deeper. We concluded above that the mass of two down quarks plus one up quark (neutron) is about 11.9 MeV/c^2. The mass of two up quarks plus one down quark (proton) is about 9.4, a difference of 2.5 MeV/c^2, which is obviously just the difference in the masses of a single down quark versus an up quark. Now this is *larger* than the difference between the masses of the neutron and proton (which is about 1.3 MeV/c^2).

We can guess what happens. If we assume that the strengths (and hence energies) of the quark–gluon and quark–quark interactions are similar for both up and down quarks, then these interactions add equally to the masses of the proton and neutron. The difference in their masses is then just the difference in the masses of the down versus up valence quarks, *offset* by a small net contribution from the electromagnetic mass of the proton.

You might be tempted to conclude that this subtle horse-trading between quantum fields and forces inside protons and neutrons is all very interesting, but ultimately irrelevant. Well, you'd be quite wrong. The simple fact that the neutron is slightly heavier than the proton underpins much of the structure of the physical world that we tend to take for granted. Put it this way, if the difference in mass was much smaller, then the proton would lose its stability and become radioactive. It would become susceptible to inverse beta-decay, transforming into a neutron with the emission of a W^+ particle and an electron-neutrino. If the difference in mass was much larger, then the fusion of protons to form helium nuclei in the centres of stars would become difficult to impossible, and no heavier elements could be formed.

Either way, the universe would be a very different place and we would certainly not be here to witness it.

This is all well and good, but it raises one final question. If the relative stability of the proton and neutron—and hence our very existence—depends on the difference in the masses of the up and down quarks, then what is the origin of this difference? *Why is the down quark heavier than the up quark?*

Recall from Chapter 14 that all elementary particles, including matter particles such as quarks and electrons, derive their mass through interactions with the Higgs field. In the case of matter particles (which are all fermions, remember) the interaction is referred to as a *Yukawa interaction*, named for Japanese theorist Hideki Yukawa. But the principle is the same. Particles that would otherwise be massless, 'two-dimensional', and moving at the speed of light interact with the Higgs field, 'absorb' a Higgs boson, gain a third dimension, and slow down. The interaction results in the appearance of terms related to $m^2\phi^2$ in the equations, and the particles 'gain mass'.

Our final question then becomes: why does the down quark interact (or 'couple') more strongly with the Higgs field when compared with the up quark? It is here that we reach the limit of our present understanding. The Higgs mechanism is a fundamental component of the current standard model of particle physics and the recent discovery of the Higgs boson tells us that it is likely to be correct. But on the question of the relative strengths of the interactions of different matter and force particles with the Higgs field the model is stubbornly silent.

It may well be that as we learn more about the Higgs boson (now that we know we can produce it at the LHC) we will discover more of its secrets. But any revelations must await us in the future.

John Wheeler used the phrase 'mass without mass' to describe the effects of superpositions of gravitational waves which could concentrate and localize energy such that a black hole is created. If this were to happen, it would mean that a black hole—the ultimate manifestation of super-high-density matter—had been created not from the matter in a collapsing star but from fluctuations in spacetime. What Wheeler really meant was that this would be a case of creating a black hole (mass) from gravitational energy.

But Wheeler's phrase is more than appropriate here. Frank Wilczek, one of the architects of QCD, used it in connection with his discussion of the results of the QCD-lite calculations in a couple of papers published in 2002[7] and the MIT Physics Annual published in 2003.[8] If much of the mass of a proton and neutron comes from the energy of interactions taking place inside these particles, then this is indeed 'mass without mass', meaning that we get the *behaviour* we tend to ascribe to mass without the need for mass as a *property*.

Does this sound familiar? Recall that in Einstein's seminal addendum to his 1905 paper on special relativity the equation he derived is actually $m = E/c^2$. This is the great insight (*not $E = mc^2$*). And Einstein was surely prescient when he wrote: 'the mass of a body is a measure of its energy content'.[9] Indeed, it is. In his book *The Lightness of Being*, Wilczek wrote:[10]

> If the body is a human body, whose mass overwhelmingly arises from the protons and neutrons it contains, the answer is now clear and decisive. The inertia of that body, with 95% accuracy, *is* its energy content.

In the fission of a U-235 nucleus, some of the energy of the colour fields inside its protons and neutrons is released, with potentially explosive consequences. In the proton–proton chain involving the fusion of four protons, the conversion of two up quarks into two down quarks, forming two neutrons in the process, results in the release of a little excess energy from its colour fields. Mass does not convert to energy. Energy is instead passed from one kind of quantum field to another.

Where does this leave us? We've certainly come a long way since the ancient Greek atomists speculated about the nature of material substance, 2,500 years ago. But for much of this time we've held to the conviction that matter is a fundamental part of our physical universe. We've been convinced that it is matter that has energy. And, although matter may be reducible to microscopic constituents, for a long time we believed that these would still be recognizable as matter—they would still possess the primary quality of mass.

Modern physics teaches us something rather different, and deeply counter-intuitive. As we worked our way ever inwards—matter into atoms, atoms into sub-atomic particles, sub-atomic particles into quantum fields and forces—we lost sight of matter completely. Matter lost its tangibility. It lost its primacy as mass became a secondary quality, the result of interactions between intangible quantum fields. What we recognize as mass is a behaviour of these quantum fields; it is not a property that belongs or is necessarily intrinsic to them.

Despite the fact that our physical world is filled with hard and heavy things, it is instead the *energy* of quantum fields that reigns supreme. Mass becomes simply a physical manifestation of that energy, rather than the other way around.

This is conceptually quite shocking, but at the same time extraordinarily appealing. The great unifying feature of the universe is the energy of quantum fields, not hard, impenetrable atoms. Perhaps this is not quite the dream that philosophers might have held fast to, but a dream nevertheless.

Five things we learned

1. We've come a long way in 2,500 years. The 'round' atoms of water favoured by Lucretius were replaced by the atoms of the chemical elements hydrogen and oxygen, in the combination H_2O.

2. We trace the mass of a cube of water ice to the nuclei of all the atoms of hydrogen and oxygen it contains, and thence to the masses of all the protons and neutrons in these nuclei. But then we run into a problem. We now know that protons and neutrons are composed of up and down quarks, but the masses of these elementary particles account for only one per cent of their host particles.

3. Sophisticated QCD calculations help to explain what's going on. Most of the mass of protons and neutrons is derived from the *energies* of interactions between quarks and massless gluons and between quarks and other quarks which take place inside these particles.

4. In addition, the masses of the quarks are derived from interactions between otherwise massless particles and the Higgs field. We conclude that mass is a secondary quality. This is 'mass without mass': we get the *behaviour* we ascribe to mass without the *property* of mass.

5. We celebrate Einstein's great prescience when, in 1905, he wrote that the mass of a body is a measure of its energy content, $m = E/c^2$.

EPILOGUE

Without the belief that it is possible to grasp reality with our theoretical constructions, without the belief in the inner harmony of our world, there could be no science. This belief is and always will remain the fundamental motive for all scientific creation.

Albert Einstein[1]

Everything in our tangible reality would appear to be constructed from intangible phantoms. I don't think it's unreasonable to declare that never in the entire history of science has the rug of understanding been pulled so firmly from beneath our feet. In *Concepts of Mass in Contemporary Physics and Philosophy*, the Israeli physicist and philosopher Max Jammer concluded: '…in spite of all the strenuous efforts of physicists and philosophers, the notion of mass, although fundamental in physics, is…shrouded in mystery'.[2]

But, of course, this is not the end. Although it's very unlikely that the way we interpret the nature of matter and the property of mass will get any easier to comprehend anytime soon, we can be reasonably assured that we don't yet have the whole story. The standard model of particle physics is an extraordinary achievement, but it is also riddled with explanatory holes. There's an awful lot it can't tell us about how the physical world is put together.

Where do we go next? The theorists have had no real alternative but to speculate. Over the past forty years or so their attempts to solve some of the standard model's more pressing problems have led them to hypotheses based on, among other things, supersymmetry and superstrings. As you read through Chapter 15, did you wonder why there was no discussion of a *prediction* for the mass of the Higgs boson *before* it was discovered in July 2012? After all, Weinberg predicted the masses of the W and Z bosons in 1967, sixteen years before these particles were discovered at CERN, so why didn't theorists do the same for the Higgs?

The standard quantum-theoretical approach to calculating the mass of the Higgs involves computing so-called radiative corrections to the particle's 'bare' mass, thereby renormalizing it. But this calculation proved to be beyond the theorists, which is why nobody really knew what the mass of the Higgs boson should be before the search for it began.

The radiative corrections involve taking account of all the different processes that a Higgs particle can undergo as it moves from place to place. These include virtual processes, involving the production of other particles and their anti-particles for a short time before these recombine. Now, the Higgs boson is obliged to couple to other particles in direct proportion to their masses, so virtual processes involving very heavy particles such as the top quark are expected to make significant contributions to the 'dressed' mass of the Higgs.

To cut a long story short, on this basis the mass of the Higgs is expected to mushroom to a size that is physically quite unrealistic. Clearly, something must be happening to cancel out the contributions from all these radiative corrections, 'tuning' the Higgs mass to a value of about 125 GeV/c^2.

There are some fairly obvious potential explanations. As American physicist Stephen Martin explained in 2011: 'The systematic cancellation of the dangerous contributions to [the Higgs mass] can only be bought about by the type of conspiracy that is better known to physicists as a symmetry.'[3]

The symmetry in question is called *supersymmetry*. Theories based on supersymmetry assume there exists a fundamental symmetry between fermions and bosons. Such theories inevitably proliferate more particles. For every fermion, the theory predicts a corresponding supersymmetric fermion (called a sfermion), which is actually a boson. So, for every particle in the standard model, the theory requires a supersymmetric partner. The partner of the electron is called the selectron (a shortening of 'scalar'-electron). Each quark is partnered by a corresponding squark.

Likewise, for every boson in the standard model, there is a corresponding supersymmetric boson, called a bosino, which is actually a fermion. Supersymmetric partners of the photon, W, and Z particles are the photino, wino, and zino.* Could supersymmetric particles (or 'sparticles') be candidates for dark matter? Several theorists believe so.

Now, supersymmetry is an assumption (or hypothesis) without any kind of foundation in empirical data, other than the observation that something like this might provide a handy mechanism to stabilize the Higgs mass. Of course, there are precedents in the history of physics for this kind of reasoning. The existence of matter and anti-matter particles reflects just such a symmetry in nature. But the symmetry between matter and anti-matter is 'exact'—aside from their different electric charges an electron and its positively charged partner, the positron, behave in much the same way and have precisely the same mass. This cannot be the case for supersymmetry. If the supersymmetry were exact, then selectrons (to take one example) could be expected to be fairly ubiquitous, or at least as common as positrons.

The simple fact that no sparticles have ever been observed means that the supersymmetry must be broken, pushing the mass range of the partners beyond reach of successive generations of

* Wino is pronounced 'weeno', presumably to avoid confusion.

particle colliders. It's fair to say that we don't have a very good theoretical explanation for how this is supposed to work, and supersymmetry theories have many, many more adjustable parameters than the standard model they are trying to fix. Some theorists haven't helped their cause by regularly 'predicting' the masses of the lightest sparticles, only to revise their calculations and make them heavier (and so further out of reach) when particle collider experiments don't find them.

On the surface, the extension of quantum chromodynamics (QCD) into string theory appears entirely logical and even, perhaps, inevitable. Instead of the mathematically unruly point particles that in QCD are arranged in a lattice and connected together by gluon fields, we introduce 'strings', two-dimensional filaments of energy. Elementary particles are then interpreted as fundamental vibrations of these strings.

This seems like a perfectly straightforward extension of existing quantum field theories, but it comes with a hefty price tag. Strings need a lot more than three spatial dimensions to vibrate in, demanding 'hidden' dimensions curled up and tucked away so tightly that they cannot be directly perceived. The strings must also be supersymmetric (hence 'superstrings'), dragging in all the caveats associated with supersymmetry theories. And, as there appear to be at least 10^{500} different ways of curling up the hidden dimensions with no clue as to the shape that nature adopted in our own universe, the theory's ability to make testable predictions is limited to non-existent.

You might then wonder why a good many theorists persist with it. There are a number of reasons, summarized by theorist-turned-philosopher Richard Dawid in his 2013 book *String Theory and the Scientific Method*. But the most compelling reason is, arguably, that—quite out of the blue—it was discovered that string theory accommodates a massless boson with a spin quantum number *s* equal to 2. This is just the kind of particle believed (at least in quantum field theories) to be required to carry the force of gravity.

For a time, it was thought that string theory offered the promise that it could accommodate gravity in some kind of final 'theory of everything'. Alas, things haven't worked out this way (despite what you might have gathered from many popular science books, magazine articles, radio, and television programmes). In truth, string theory is not actually a 'theory'. At best it is a 'framework' or series of connected hypotheses. String theorists have made much progress in establishing some consistency between its structures, but they have now largely given up declaring that this is a potential theory of everything.

However, they retain some hopes for a *quantum theory of gravity*. In this regard, and despite what many string theorists might wish to claim, string theory is not the only 'game in town'. It is certainly not the only candidate for a quantum theory of gravity. Think of it this way. String theory approaches gravity from the perspective of a string-based quantum field theory. An alternative approach is to start with the general theory of relativity and try to find a way to 'quantize' it. One potential alternative solution that results from this approach is called loop quantum gravity, or LQG, which has involved many theorists but is perhaps most closely associated with Lee Smolin and Carlo Rovelli.

In a string quantum gravity, gravity is the result of quantum excitations or vibrations of strings in an essentially classical spacetime 'container'. In LQG, gravity is not a 'force'; it is a manifestation of the quantum nature of spacetime geometry, literally of 'atoms' of space and time. In principle it requires only four spacetime dimensions and does not require supersymmetry.

In LQG, it is *geometry* that is quantized. An area of space is composed of an integral number of fundamental quanta with dimensions about the square of the 'Planck length', an unimaginably small 1.6×10^{-35} metres.[4] The quanta of volume are related to the cube of the Planck length.

In *Three Roads to Quantum Gravity*, first published in 2000, Smolin stuck his neck out and suggested that we would have 'the basic framework of the quantum theory of gravity by 2010, 2015 at the outside'.[5] He went on to suggest that we would be teaching the theory of quantum gravity to high-school students by the end of the twenty-first century.

An assessment by Rovelli in 2012 suggested that, after twenty-five years of theoretical effort, LQG has broadly delivered or even over-delivered on its early promise. Many of its conceptual and mathematical problems have now been resolved and the result is a consistent quantum field theory whose classical limit is general relativity.

The stubborn problem that remains concerns the theory's ability to be put to the test. As the theory applies to physics at the 'Planck scale', this is necessarily physics that is likely to remain firmly out of reach. It's possible that LQG might make contact with empirical data, for example through quantum effects in the universe before the onset of cosmic inflation, which may have left their imprints in the cosmic background radiation. Whether such subtle effects will be sufficient to differentiate LQG from alternative explanations remains to be seen. As Rovelli says: '…the situation in quantum gravity is in my opinion…far better than twenty-five years ago, and, one day out of two, I am optimistic.'[6]

If LQG or something very like it ultimately proves to be a viable quantum theory of gravity, we will certainly have cause for celebration, but we should be under no illusions. The ambition of the theorists engaged in LQG is to find a quantum description of general relativity. This will not solve—it is not designed to solve—the present problems with the standard model of particle physics.

> The philosophy underlying loop gravity is that we are not near the end of physics, we better not dream of a final theory of everything, and we better solve one problem at [a] time, which is hard enough.[7]

Let's use these last words to pause for a moment and reflect on the 2,500-year history of physics since the days of the Greek atomists. I don't know about you, but it seems to me that the great revolutions in scientific thinking that have shaped the way we seek to comprehend the universe have involved some pretty dramatic changes in *perspective*. Each revolution has profoundly changed the way we think about the 'fabric', the way we interpret a physical reality constructed from its most basic ingredients: space, time, matter, and energy.

Such changes in perspective have proved to be extraordinarily powerful. They have yielded insights that have allowed us to manipulate our reality, largely (though certainly not exclusively) to our advantage. But being able to do things with reality doesn't mean we've really understood it. We learn from the history of the fabric that concepts such as mass, though entirely familiar, have *never* been properly understood.

Modern science has revealed the extraordinarily rich structure of our empirical reality, a reality consisting of things-as-they-appear and things-as-they-are-measured. But it would be naïve to think that this extraordinary success has drawn us ever closer to comprehending a reality of things-in-themselves. If anything, the richness we have discovered would seem to have dragged us further away. To paraphrase the philosopher Bernard d'Espagnat, our understanding of the basic structure of physical reality 'is an "ideal" from which we remain distant. Indeed, a comparison with conditions that ruled in the past suggests that we are a great deal more distant from it than our predecessors thought they were a century ago."[8]

I find this truly exciting. Physicists at the end of the nineteenth century famously thought they'd got it all figured out.[9] Now we know better. We've since learned an awful lot, but we're also very aware of what we don't know and can't explain. If we dream of a destination where we have ultimate knowledge of everything, then I doubt we'll ever get there. But I'm also convinced there'll be much to see—and much still to learn and enjoy—as we make the journey.

ENDNOTES

PREFACE

1. Paul Dirac, *Nature*, **126**, 1930, pp. 605–6, quoted in Helge Kragh, *Dirac: A Scientific Biography*, Cambridge University Press, 1990, p. 97.
2. Stephen Hawking, *A Brief History of Time: From the Big Bang to Black Holes*, Bantam Press, 1988, p. vi.

CHAPTER 1: THE QUIET CITADEL

1. Lucretius, *On the Nature of the Universe*, trans. R.E. Latham, Penguin Books, London (first published 1951), pp. 61–2.
2. Recent X-ray studies have revealed that these papyri were written using metal-based inks, contradicting previous wisdom and offering prospects for optimizing computer-aided tomography of unrolled scrolls. See Emmanuel Brun, Marine Cotte, Jonathan Wright, *et al.*, *Proceedings of the National Academy of Sciences*, **113** (2016), pp. 3751–4.
3. Epicurus wrote: 'To begin with, nothing comes into being out of what is non-existent.' Epicurus, letter to Herodotus, reproduced in Diogenes Laërtius, *Lives of the Eminent Philosophers*, Book X, **38**, trans. Robert Drew Hicks, Loeb Classical Library (1925), http://en.wikisource.org/wiki/Lives_of_the_Eminent_Philosophers.
4. Or as Lucretius put it: 'The second great principle is this: *nature resolves everything into its component atoms and never reduces anything to nothing.*' Lucretius, *On the Nature of the Universe* (n 1), p. 33.
5. As Epicurus explains: 'And if there were no space (which we call also void and place and intangible nature), bodies would have nothing in which to be and through which to move, as they are plainly seen to move. Beyond bodies and space there is nothing which by mental apprehension or on its analogy we can conceive to exist.' Epicurus, letter to Herodotus (n 3), Book X, **40**.
6. Lucretius wrote: 'Granted that the particles of matter are absolutely solid, we can still explain the composition and behaviour of soft

things—air, water, earth, fire—by their intermixture with empty space.' Lucretius, *On the Nature of the Universe* (n 1), p. 44.

7. See Plato, *Timaeus and Critias*, Penguin, London (1971), pp. 73–87. Plato built air, fire, and water from one type of triangle and earth from another. Consequently, Plato argued that it was not possible to transform earth into other elements.

8. Lucretius wrote: 'It can be shown that Neptune's bitter brine results from a mixture of rougher atoms with smooth. There is a way of separating the two ingredients and viewing them in isolation by filtering the sweet fluid through many layers of earth so that it flows out into a pit and loses its tang. It leaves behind the atoms of unpalatable brine because owing to their roughness they are more apt to stick fast in the earth.' Lucretius, *On the Nature of the Universe* (n 1), pp. 73–4.

9. In *The Metaphysics*, Aristotle wrote: 'They [Leucippus and Plato] maintain that motion is always in existence: but why, and in what way, they do not state, nor how is this the case; nor do they assign the cause of this perpetuity of motion.' Aristotle, *The Metaphysics*, Book XII, **1071b**, trans. John H. McMahon, Prometheus Books, New York (1991), p. 256.

10. Lucretius wrote: 'Since the atoms are moving freely through the void they must all be kept in motion either by their own weight or on occasion by the impact of another atom.' Lucretius, *On the Nature of the Universe* (n 1), p. 62.

11. Lucretius, *On the Nature of the Universe* (n 1), p. 66.

12. Lucretius again: 'Indeed, even visible objects, when set at a distance, often disguise their movements. Often on a hillside fleecy sheep, as they crop their lush pasture, creep slowly onward, lured this way or that by grass that sparkles with fresh dew, while full-fed lambs gaily frisk and butt. And yet, when we gaze from a distance, we see only a blur—a white patch stationary on the green hillside.' Lucretius, *On the Nature of the Universe* (n 1), p. 69.

13. Lucretius concluded: 'It follows that nature works through the agency of invisible bodies.' Lucretius, *On the Nature of the Universe* (n 1), p. 37.

14. Lucretius certainly thought so: 'Observe what happens when sunbeams are admitted into a building and shed light on its shadowy places. You will see a multitude of tiny particles mingling in a

multitude of ways in the empty space within the light of the beam, as though contending in everlasting conflict...their dancing [of the particles in the sunbeam] is an actual indication of underlying movements of matter that are hidden from our sight....You must understand that they all derive this restlessness from the atoms.' Lucretius, *On the Nature of the Universe* (n 1), pp. 63–4.

15. Democritus, in Hermann Diels, *Die Fragmente von Vorsokratiker*, Weidmann, Berlin (1903), **117**, p. 426. The German translation is given as: 'In Wirklichkeit wissen wir nichts; denn die Wahrheit liegt in der Tiefe.' The quoted English translation is taken from Samuel Sambursky, *The Physical World of the Greeks*, 2nd edn, Routledge, London (1960), p. 131.

CHAPTER 2: THINGS-IN-THEMSELVES

1. Immanuel Kant, *Critique of Pure Reason* (see Sebastian Gardner, *Kant and the Critique of Pure Reason*, Routledge, Abingdon, 1999, p. 205).

2. The authors of the entry on medieval philosophy in the online *Stanford Encyclopedia of Philosophy* put it rather succinctly: 'Here is a recipe for producing medieval philosophy: Combine classical pagan philosophy, mainly Greek but also in its Roman versions, with the new Christian religion. Season with a variety of flavorings from the Jewish and Islamic intellectual heritages. Stir and simmer for 1300 years or more, until done.' Paul Vincent Spade, Gyula Klima, Jack Zupko, and Thomas Williams, 'Medieval Philosophy', *Stanford Encyclopedia of Philosophy*, Spring 2013, p. 4.

3. 'We should then be never required to try our strength in contests about the soul with philosophers, those patriarchs of heretics, as they may be fairly called.' Tertullian, *De Anima*, Chapter III, translated by Peter Holmes, http://www.earlychristianwritings.com/text/tertullian10.html.

4. Boyle wrote: 'To convert Infidels to the Christian Religion is a work of great Charity and kindnes to men', in J.J. MacIntosh (ed.), *Boyle on Atheism*, University of Toronto Press (2005), p. 301.

5. René Descartes, *Discourse on Method and The Meditations*, trans. F.E. Sutcliffe, Penguin, London (1968), p. 53.

6. Descartes wrote: 'And indeed, from the fact that I perceive different sorts of colours, smells, tastes, sounds, heat, hardness, etc., I rightly conclude that there are in the bodies from which all these diverse perceptions of the senses come, certain varieties corresponding to them, although perhaps these varieties are not in fact like them.' Descartes, ibid., p. 159.

7. Locke wrote: 'For division (which is all that a mill, or pestle, or any other body, does upon another, in reducing it to insensible parts) can never take away either solidity, extension, figure, or mobility from any body.... These I call original or *primary qualities*.... Secondly, such qualities which in truth are nothing in the objects themselves but powers to produce various sensations in us by their primary qualities, i.e. by the bulk, figure, texture, and motion of their insensible parts, as colours, sounds, tastes, &c. These I call *secondary qualities*.' John Locke, *Essay Concerning Human Understanding*, Book II, Chapter VIII, sections 9 and 10 (first published 1689), ebooks@Adelaide, The University of Adelaide, https://ebooks.adelaide.edu.au/l/locke/john/l81u/index.html.

8. Berkeley wrote: 'They who assert that figure, motion, and the rest of the primary or original qualities do exist without the mind...do at the same time acknowledge that colours, sounds, heat cold, and suchlike secondary qualities, do not.... For my own part, I see evidently that it is not in my power to frame an idea of a body extended and moving, but I must withal give it some colour or other sensible quality which is acknowledged to exist only in the mind. In short, extension, figure, and motion, abstracted from all other qualities, are inconceivable. Where therefore the other sensible qualities are, there must these be also, to wit, in the mind and nowhere else.' George Berkeley, *A Treatise Concerning the Principles of Human Knowledge*, The Treatise para. 10 (first published 1710), ebooks@Adelaide, The University of Adelaide, https://ebooks.adelaide.edu.au/b/berkeley/george/b51tr/index.html.

CHAPTER 3: AN IMPRESSION OF FORCE

1. The full quotation reads: '*The alteration of motion is ever proportional to the motive force impressed; and is made in the direction of the right line in which that force is impressed....* And this motion (being always directed the

same way with the generating force), if the body moved before, is added to or subducted [i.e. subtracted] from the former motion, according as they directly conspire with or are directly contrary to each other; or obliquely joined, when they are oblique, so as to produce a new motion compounded from the determination of both.' Isaac Newton, *Mathematical Principles of Natural Philosophy*, first American edition trans. Andrew Motte, Daniel Adee, New York, 1845, p. 83.

2. See, e.g., Catherine Wilson, *Epicureanism at the Origins of Modernity*, Oxford University Press, 2008, especially pp. 51–5.

3. Newton wrote: 'The extension, hardness, impenetrability, mobility, and *vis inertia* of the whole, result from the extension, hardness, impenetrability, mobility, and *vires inertia* of the parts; and thence we conclude the least particles of all bodies to be also all extended, and hard and impenetrable, and moveable, and endowed with their proper *vires inertia*.' Newton, *Mathematical Principles* (n 1), p. 385. In this quotation, *vis inertia* simply means 'inertia', the measure of the resistance of a body to acceleration ('vis' means force or power). Newton then equates the inertia of an object to the sum total—the *vires inertia*—of all its constituent atoms ('vires' is the plural of 'vis').

4. Newton, ibid., p. 73.

5. The first law is given in the *Mathematical Principles* as: '*Every body perseveres in its state of rest, or of uniform motion in a right line [straight line], unless it is compelled to change that state by forces impressed thereon.*' Newton, ibid., p. 83.

6. Newton wrote: '*An impressed force is an action exerted upon a body, in order to change its state, either of rest, or of moving forward uniformly in a right line. This force consists in the action only; and remains no longer in the body, when the action is over. For a body maintains every new state it acquires, by its vis inertiae only. Impressed forces are of different origins as from percussion, from pressure, from centripetal force.*' Newton, ibid., p. 74.

7. Newton, ibid., p. 83.

8. Suppose we apply the force for a short amount of time, Δt.* This effects a change in the linear momentum by an amount $\Delta(mv)$. If we

---

* We use the Greek symbol delta (Δ) to denote 'difference'. So, if we apply the force at some starting time t_1 and remove it a short time later, t_2, then Δt is equal to the time difference, t_2-t_1.

now assume that the inertial mass m is intrinsic to the object and does not change with time or with the application of the force (which seems entirely reasonable and justified), then the change in linear momentum is really the inertial mass multiplied by a change in velocity: $\Delta(mv) = m\Delta v$. Applying the force may change the magnitude of the velocity (up or down) and it may change the direction in which the object is moving. Newton's second law is then expressed mathematically as $F\Delta t = m\Delta v$: impressing the force for a short time changes the velocity (and hence linear momentum) of the object. Now this equation may not look very familiar. But we can take a further step: dividing both sides of this equation by Δt gives $F = m\Delta v/\Delta t$. The ratio $\Delta v/\Delta t$ is the rate of change of velocity (magnitude *and* direction) with time. We have another name for this quantity: it is called *acceleration*, usually given the symbol a. Hence Newton's second law can be re-stated as the much more familiar $F = ma$, or Force equals Inertial mass × Acceleration.

9. Newton states the third law thus: '*To every action there is always opposed an equal reaction: or the mutual actions of two bodies upon each other are always equal, and directed to contrary parts.*' Newton, *Mathematical Principles* (n 1), p. 83.

10. Mach wrote: 'With regard to the concept of "mass", it is to be observed that the formulation of Newton, which defines mass to be the quantity of matter of a body as measured by the product of its volume and density, is unfortunate. As we can only define density as the mass of a unit of volume, the circle is manifest.' Ernst Mach, *The Science of Mechanics*, quoted in Max Jammer, *Concepts of Mass in Contemporary Physics and Philosophy*, Princeton University Press, 2000, p. 11.

11. These quotations are derived from H.W. Trumbull, J.F. Scott, A.R. Hall, and L. Tilling (eds), *The Correspondence of Isaac Newton*, Cambridge University Press, Vol. 2, pp. 437–9, and are quoted in Lisa Jardine, *The Curious Life of Robert Hooke: The Man Who Measured London*, Harper Collins, London, 2003, p. 6.

12. Newton wrote: '*That the fixed stars being at rest, the periodic times of the five primary planets, and (whether of the sun about the earth or) of the earth about the sun, are in the sesquiplicate proportion of their mean distances from the sun.*' Newton, *Mathematical Principles* (n 1), p. 388. 'Sesquiplicate' means 'raised to the power $\frac{3}{2}$'.

13. Only six planets were known in Newton's time, but we can now extend Kepler's logic to eight planets as follows:

Planet	Orbital period T (days)	Orbital radius r (10⁶ km)	T^2/r^3
Mercury	87.97	57.91	0.1996
Venus	224.70	108.21	0.1996
Earth	356.26	149.51	0.1949
Mars	686.96	227.94	0.1996
Jupiter	4,332.59	778.57	0.1994
Saturn	10,579.22	1,433.45	0.1949
Uranus	30,799.10	2,876.68	0.1996
Neptune	60,193.03	4,503.44	0.1992

For sure, there is some variation in the ratio T^2/r^3, but the deviation from the mean value of 0.1984 is never greater than two per cent.

14. Proposition VII states: 'That there is a power of gravity tending to all bodies, proportional to the several quantities of matter which they contain.' Newton, *Mathematical Principles* (n 1), p. 397.

15. Proposition VIII states: 'In two spheres mutually gravitating each towards the other, if the matter in places on all sides round about and equi-distant from the centres is similar, the weight of either sphere towards the other will be reciprocally as the square of the distance between their centres.' Newton, *Mathematical Principles* (n 1), p. 398.

16. Newton wrote: 'Hitherto we have explained the phænomena of the heavens and of our sea, by the power of Gravity, but have not yet assigned the cause of this power....I have not been able to discover the cause of those properties of gravity from phænomena, and I frame no hypotheses.' Newton, *Mathematical Principles* (n 1), p. 506.

CHAPTER 4: THE SCEPTICAL CHYMISTS

1. Stanislao Cannizzaro, *Sketch of a Course of Chemical Philosophy*, in *Il Nuovo Cimento*, 7 (1858), pp. 321–66. Reproduced as Alembic Club (Edinburgh) Reprint 18, The University of Chicago Press, 1911. This quote appears on p. 11.

2. See, e.g., Jed Z. Buchwald, *The Rise of the Wave Theory of Light*, University of Chicago Press, 1989, pp. 6–7.

3. Isaac Newton, *Opticks*, 4th edn (first published 1730), Dover Books, New York, 1952. This quote from Querie 29 appears on p. 370.

4. Querie 31 contains the passage: 'Have not the Small Particles of Bodies certain Powers, Virtues or Forces, by which they act at a distance, not only upon the Rays of Light…but also upon one another for producing a great Part of the Phænomena of Nature? For it's well known that Bodies act upon one another by the attractions of Gravity, Magnetism and Electricity…and make it not improbable but that there may be more attractive Powers than these.' Newton, *Opticks*, ibid., pp. 375–6.

5. Querie 31 continues: 'I had rather infer…that their Particles attract one another by some Force, which in immediate Contact is exceeding strong, at small distances performs the chymical Operations above-mention'd, and reaches not far from the Particles with any sensible Effect.' Newton, *Opticks*, ibid., p. 389.

6. Boyle wrote: 'And, to prevent mistakes, I must advertise you, that I now mean by elements, and those chymists, that speak plainest, do by their principles, certain primitive and simple, or perfectly unmingled bodies, which not being made of any other bodies, or of one another, are the ingredients, of which all those called perfectly mixt bodies are immediately compounded, and into which they are ultimately resolved.' Robert Boyle, *The Sceptical Chymist*, reproduced in Thomas Birch, *The Works of the Honourable Robert Boyle*, Vol. 1, London, 1762, p. 562.

7. Joseph Priestley, *Experiments and Observations on Different Kinds of Air*, Vol. II, Section III, London 1775, http://web.lemoyne.edu/~GIUNTA/priestley.html.

8. Dalton wrote: 'An enquiry into the relative weights of the ultimate particles of bodies is a subject, as far as I know, entirely new: I have already been prosecuting this enquiry with remarkable success. The principle cannot be entered upon in this paper; but I shall just subjoin the results, as they appear to be ascertained by my experiments.' The paper was published two years later, in 1805. John Dalton, quoted in Frank Greenaway, *John Dalton and the Atom*, Heinemann, London, 1966, p. 165.

9. Cannizzaro, *Sketch of a Course of Chemical Philosophy* (n 1), p. 11.

10. Cannizzaro, *Sketch of a Course of Chemical Philosophy* (n 1), p. 11.

11. Cannizzaro, *Sketch of a Course of Chemical Philosophy* (n 1), p. 12.

12. Einstein wrote: 'In this paper it will be shown that, according to the molecular-kinetic theory of heat, bodies of a microscopically visible size suspended in liquids must, as a result of thermal molecular motions, perform motions of such magnitude that they can be easily observed with a microscope. It is possible that the motions to be discussed here are identical with so-called Brownian molecular motion; however the data available to me on the latter are so imprecise that I could not form a judgement on the question.' Albert Einstein, *Annalen der Physik*, **17** (1905), pp. 549–60. This paper is translated and reproduced in John Stachel (ed.), *Einstein's Miraculous Year: Five Papers that Changed the Face of Physics*, Centenary edn, Princeton University Press, 2005. The quote appears on p. 85.

CHAPTER 5: A VERY INTERESTING CONCLUSION

1. Albert Einstein, *Annalen der Physik*, **18** (1905), pp. 639–41, trans. and repr. in John Stachel (ed.), *Einstein's Miraculous Year: Five Papers that Changed the Face of Physics*, centenary edn, Princeton University Press, 2005. The quote appears on p. 164.

2. Maxwell wrote that the speed is: '...so nearly that of light, that it seems we have strong reason to conclude that light itself (including radiant heat, and other radiations if any) is an electromagnetic disturbance in the form of waves propagated through the electromagnetic field according to electromagnetic laws.' James Clerk Maxwell, *A Dynamical Theory of the Electromagnetic Field*, Part I, §20 (1864), https://en.wikisource.org/wiki/A_Dynamical_Theory_of_the_Electromagnetic_Field/Part_I.

3. Einstein wrote: 'The introduction of a "light ether" will prove to be superfluous, inasmuch as the view to be developed here will not require a "space at absolute rest" endowed with special properties...'. Albert Einstein, *Annalen der Physik*, **17** (1905), pp. 891–921, trans. and repr. in Stachel, *Einstein's Miraculous Year* (n 1), p. 124.

4. We make our first set of measurements whilst the train is stationary. The light travels straight up and down, travelling a total distance of

$2d_0$, where d_0 is the height of the carriage, let's say in a time t_0. We therefore know that $2d_0 = ct_0$, where c is the speed of light, as the light travels up (d_0) and then down (another d_0) in the time t_0 at the speed c. If we know d_0 precisely, we could use this measurement to determine the value of c. Alternatively, if we know c we can determine d_0. We now step off the train and repeat the measurement as the train moves past us with velocity v, where v is a substantial fraction of the speed of light. From our perspective on the platform the light path looks like 'Λ'. Let's assume that total time required for the light to travel this path is t. If we join the two ends of the Λ together we form an equilateral triangle, Δ. We know that the base of this triangle has a length given by vt, the distance the train has moved forward in the time t. The other two sides of the triangle measure a distance d and we know that $2d = ct$ (remember that the speed of light is assumed to be constant). If we now draw a perpendicular (which will have length equal to d_0) from the apex of the triangle and which bisects the base, we can use Pythagoras' theorem: the square of the hypotenuse is equal to the sum of the squares of the other two sides, or $d^2 = d_0{}^2 + (\frac{1}{2}vt)^2$. But we know that $d = \frac{1}{2}ct$ and from our earlier measurement $d_0 = \frac{1}{2}ct_0$, so we have: $(\frac{1}{2}ct)^2 = (\frac{1}{2}ct_0)^2 + (\frac{1}{2}vt)^2$. We can now cancel all the factors of $\frac{1}{2}$ and multiply out the brackets to give: $c^2t^2 = c^2t_0{}^2 + v^2t^2$. We gather the terms in t^2 on the left-hand side and divide through by c^2 to obtain: $t^2(1 - v^2/c^2) = t_0{}^2$, or, re-arranging and taking the square-root: $t = \gamma t_0$, where $\gamma = 1/\sqrt{(1 - v^2/c^2)}$.

5. On 21 September 1908, Minkowski began his address to the 80th Assembly of German Natural Scientists and Physicians with these words: 'The views of space and time which I wish to lay before you have sprung from the soil of experimental physics, and therein lies their strength. They are radical. Henceforth space by itself, and time by itself, are doomed to fade away into mere shadows, and only a kind of union of the two will preserve an independent reality.' Hermann Minkowski, 'Space and Time' in Hendrik A. Lorentz, Albert Einstein, Hermann Minkowski, and Hermann Weyl, *The Principle of Relativity: A Collection of Original Memoirs on the Special and General Theory of Relativity*, Dover, New York, 1952, p. 75.

6. Albert Einstein, *Annalen der Physik*, **18** (1905), pp. 639–41, in Stachel, *Einstein's Miraculous Year* (n 1), p. 161.

7. The energy carried away by the light bursts in the stationary frame of reference is E, compared with γE in the moving frame of reference. Energy must be conserved, so we conclude that the difference must be derived from the kinetic energies of the object in the two frames of reference. This difference in kinetic energy must be, therefore, $\gamma E - E$, or $E(\gamma - 1)$. As it stands, the term $E(\gamma - 1)$ isn't very informative, and the Lorentz factor γ – which, remember, equals $1/\sqrt{(1 - v^2/c^2)}$— is rather cumbersome. But we can employ a trick often used by mathematicians and physicists. Complex functions like γ can be recast as the sum of an infinite series of simpler terms, called a Taylor series (for eighteenth-century English mathematician Brook Taylor). The good news is that for many complex functions we can simply look up the relevant Taylor series. Even better, we often find that the first two or three terms in the series provide an approximation to the function that is good enough for most practical purposes. The Taylor series in question is: $1/\sqrt{(1 + x)} = 1 - (1/2)x + (3/8)x^2 - (5/16)x^3 + (35/128)x^4 - (63/256)x^5 + \ldots$. Substituting $x = -v^2/c^2$ means that terms in x^2 are actually terms in v^4/c^4 and so on for higher powers of x. Einstein was happy to leave these out, writing: 'Neglecting magnitudes of the fourth and higher order...'. Neglecting the higher-order terms gives rise to a small error of the order of a few per cent for speeds v up to about fifty per cent of the speed of light, but the error grows dramatically as v is further increased. So, substituting $x = -v^2/c^2$ in the first two terms in the Taylor series means that we can approximate γ as $1 + \frac{1}{2}v^2/c^2$. If we now put this into the expression for the difference in kinetic energies above, we get: $E(\gamma - 1) \sim \frac{1}{2}Ev^2/c^2$, or $\frac{1}{2}(E/c^2)v^2$. Did you see what just happened? We know that the expression for kinetic energy is $\frac{1}{2}mv^2$, and in the equation for the difference we see that the velocity v is unchanged. Instead, the energy of the light bursts comes from the *mass* of the object, which falls by an amount $m = E/c^2$.

8. Einstein, *Annalen der Physik*, **18** (1905), p. 164, in Stachel, *Einstein's Miraculous Year* (n 1).

CHAPTER 6: INCOMMENSURABLE

1. Max Jammer, *Concepts of Mass in Contemporary Physics and Philosophy*, Princeton University Press, 2000, p. 61.

2. Einstein wrote: 'It is not good to introduce the concept of the [relativistic] mass M...of a moving body for which no clear definition can be given. It is better to introduce no other mass concept than the "rest mass", m. Instead of introducing M it is better to mention the expression for the momentum and energy of a body in motion.' Albert Einstein, letter to Lincoln Barnett, 19 June 1948. A facsimile of part of this letter is reproduced, together with an English translation, in Lev Okun, *Physics Today*, June 1989, p. 12.

3. Okun wrote: '...the terms "rest mass" and "relativistic mass" are redundant and misleading. There is only one mass in physics, *m*, which does not depend on the reference frame. As soon as you reject the "relativistic mass" there is no need to call the other mass the "rest mass" and to mark it with the index 0.' In his opening paragraph, picked out in a large, bold font, he declares: 'In the modern language of relativity theory there is only one mass, the Newtonian mass *m*, which does not vary with velocity.' Okun, ibid., p. 11.

4. See, e.g., A.P. French, *Special Relativity*, MIT Introductory Physics Series, Van Nostrand Reinhold, London, 1968 (reprinted 1988), p. 23.

5. Albert Einstein, *Annalen der Physik*, **18** (1905), pp. 639–41. Trans. and repr. in John Stachel (ed.), *Einstein's Miraculous Year: Five Papers that Changed the Face of Physics*, centenary edn, Princeton University Press, 2005. The quote appears on p. 164.

6. Paul Feyerabend, 'Problems of Empiricism', in R.G. Colodny, *Beyond the Edge of Certainty*, Englewood-Cliffs, New Jersey (1965), p. 169. This quote is reproduced in Jammer, *Concepts of Mass* (n 1), p. 57.

CHAPTER 7: THE FABRIC

1. John Wheeler with Kenneth Ford, *Geons, Black Holes and Quantum Foam: A Life in Physics*, W.W. Norton & Company, New York, 1998, p. 235.

2. Albert Einstein, 'How I Created the Theory of Relativity', lecture delivered at Kyoto University, 14 December 1922, trans. Yoshimasa A. Ono, *Physics Today*, August 1982, p. 47.

3. Albert Einstein, in the 'Morgan manuscript', quoted by Abraham Pais, *Subtle is the Lord: The Science and the Life of Albert Einstein*. Oxford University Press, 1982, p. 178.

4. Wheeler with Ford, *Geons, Black Holes and Quantum Foam* (n 1).

5. Albert Einstein, quoted by Pais, ibid., p. 212.
6. Albert Einstein, letter to Heinrich Zangger, 26 November 1915. This quote is reproduced in Alice Calaprice (ed.), *The Ultimate Quotable Einstein*, Princeton University Press, 2011, p. 361.
7. We can get some sense for what the Schwarzschild solutions tell us by supposing we measure two events at some distance far away from a massive object, where the effects of spacetime curvature can be safely ignored. Spacetime here is flat, and we note the time interval Δt between the two events. The Schwarzschild solutions show that the same measurement performed closer to the object where spacetime is more curved will yield a different result, $\Delta t'$, where $\Delta t' = \Delta t/(1 - R_s/r)$. Here R_s is the *Schwarzschild radius*, given by Gm/c^2, where m is the mass of the object, c is the speed of light, and G is the gravitational constant. Let us further assume that we're well outside the Schwarzschild radius, so r is much larger than R_s. (The Schwarzschild radius of the Earth is about 9 millimetres.) This means that the term in brackets is a little smaller than 1, so $\Delta t'$ is slightly larger than Δt, or time intervals are dilated. This is *gravitational time dilation*—clocks run more slowly where the effects of gravity (spacetime curvature) are stronger. It is an effect entirely separate and distinct from the time dilation of special relativity, which is caused by making measurements in different inertial frames of reference. We can make a similar set of deductions for a measurement of radial distance interval Δr, at a fixed time such that $\Delta t = 0$. We find $\Delta r' = (1 - R_s/r)\Delta r$. Now $\Delta r'$ is slightly *smaller* than Δr. Distances contract under the influence of a gravitational field.
8. See *Metromnia*, National Physical Laboratory, **18**, Winter 2005.

CHAPTER 8: IN THE HEART OF DARKNESS

1. Albert Einstein, *Proceedings of the Prussian Academy of Sciences*, **142** (1917), quoted in Walter Isaacson, *Einstein: His Life and Universe*, Simon & Shuster, New York, 2007, p. 255.
2. Newton wrote: '...and lest the systems of the fixed stars should, by their gravity, fall on each other mutually, he hath placed those systems at immense distances one from another'. Isaac Newton, *Mathematical Principles of Natural Philosophy*, first American edn trans. Andrew Motte, Daniel Adee, New York, 1845, p. 504.

3. See, e.g., Steven Weinberg, *Cosmology*, Oxford University Press, 2008, p. 44.

4. According to Ukranian-born theoretical physicist George Gamow: 'When I was discussing cosmological problems with Einstein he remarked that the introduction of the cosmological term was the biggest blunder he ever made in his life.' George Gamow, *My World Line: An Informal Autobiography*, Viking Press, New York, 1970, p. 149, quoted in Walter Isaacson, *Einstein* (n 1), pp. 355–6.

5. Hubble's law can be expressed as $v = H_0 D$, where v is the velocity of the galaxy, H_0 is Hubble's constant as measured in the present time, and D is the so-called 'proper distance' of the galaxy measured from the Earth, such that the velocity is then given simply as the rate of change of this distance. Although it is often referred to as a 'constant', in truth the Hubble parameter H varies with time depending on assumptions regarding the rate of expansion of the universe. Despite this, the age of the universe can be roughly estimated as $1/H_0$. A value of H_0 of 67.74 kilometres per second per megaparsec (or 2.195×10^{-18} per second) gives an age for the universe of 45.66×10^{16} seconds, or 14.48 billion years. (The age as determined by the Planck satellite mission is 13.82 billion years, so the universe is a little younger than it's 'Hubble age' would suggest.)

6. Lemaître wrote: 'Everything happens as though the energy in vacuo would be different from zero.' Georges Lemaître, *Proceedings of the National Academy of Sciences*, **20** (1934), pp. 12–17, quoted in Harry Nussbaumer and Lydia Bieri, *Discovering the Expanding Universe*, Cambridge University Press, 2009, p. 171.

7. In 1922, Russian physicist and mathematician Alexander Friedmann offered a number of different solutions of Einstein's original field equations. These can be manipulated to yield a relatively simple expression for the critical density (ρ_c) of mass-energy required for a flat universe: $\rho_c = 3H_0^2/8\pi G$, where H_0 is the Hubble constant and G is the gravitational constant. The value of H_0 deduced from the most recent measurements of the cosmic background radiation is 67.74 kilometres per second per megaparsec, or 2.195×10^{-18} per second. With $G = 6.674 \times 10^{-11}$ Nm^2kg^{-2} (or $m^3kg^{-1}s^{-2}$), and $H_0^2 = 4.818 \times 10^{-36}$ s^{-2}, we get $\rho_c = 8.617 \times 10^{-27}$ kgm^{-3}, which translates to 8.617×10^{-30} gcm^{-3}. Compare this with the density of air at sea level at a temperature of 15°C, which is 0.001225 gcm^{-3}.

8. The mass of a proton is 1.67×10^{-24} g, so we need a critical density of the order of 5.16×10^{-6} protons per cubic centimetre. The length of St Paul's Cathedral is 158 metres, with a height of 111 metres and a width between transepts of 75 metres. We can combine these to obtain a rough estimate of the volume, $158 \times 111 \times 75 = 1.315 \times 10^6$ m^3, or 1.315×10^{12} cm^3. If the critical density ρ_c is equivalent to 5.16×10^{-6} protons per cm^3, then to meet this density we would need to fill St Paul's Cathedral with 6.79×10^6 protons, which I've rounded up to an average of 7 million protons. To calculate the number of protons and neutrons in the air inside St Paul's Cathedral, I've assumed the air density at sea level (see n 7) of 0.001225 gcm^{-3}. This gives the total mass of air inside the Cathedral of 1.611×10^9 g. Assuming an average proton/neutron mass of 1.6735×10^{-24} g gives a total number of protons and neutrons inside the cathedral of 9.63×10^{32}.

9. So, what is the density of dark energy? If we assume ρ_c is 8.62×10^{-30} gcm^{-3}, we know from the latest Planck satellite results that dark energy must account for about sixty-nine per cent of this, or 5.94×10^{-30} gcm^{-3}. This is really a mass density, so we convert it to an energy density using $E = mc^2$, giving 5.34×10^{-16} Jcm^{-3}. If we call this vacuum energy density ρ_v, we can use the relation $\Lambda = (8\pi G/c^4)\rho_v$ to calculate a value for the cosmological constant of 1.109×10^{-52} per square metre. We can put this dark or vacuum energy density into perspective. The chemical energy released on combustion of a litre (1,000 cubic centimetres) of petrol (gasoline for American readers) is 32.4 million joules, implying an energy density of 32,400 Jcm^{-3}. So, 'empty' space-time has an energy density about 1.6 hundredths of a billionth of a billionth (1.6×10^{-20}) of the chemical energy density of petrol. It might not be completely empty, but it's still the 'vacuum', after all.

CHAPTER 9: AN ACT OF DESPERATION

1. Max Planck, letter to Robert Williams Wood, 7 October 1931, quoted in Armin Hermann, *The Genesis of Quantum Theory (1899–1913)*, trans. Claude W. Nash, MIT Press, Cambridge, MA, 1971, p. 21.

2. Max Planck, letter to Wilhelm Ostwald, 1 July 1893, quoted in J.L. Heilbron, *The Dilemmas of an Upright Man: Max Planck and the Fortunes of German Science*, Harvard University Press, 1996, p. 15.

3. Max Planck, *Physikalische Abhandlungen und Vorträge*, Vol. 1, Vieweg, Braunschweig, 1958, p. 163, quoted in Heilbron, ibid., p. 14.

4. Einstein wrote: 'If monochromatic radiation (of sufficiently low density) behaves...as though the radiation were a discontinuous medium consisting of energy quanta of magnitude [$h\nu$], then it seems reasonable to investigate whether the laws governing the emission and transformation of light are also constructed such as if *light consisted of such energy quanta.*' Albert Einstein, *Annalen der Physik*, **17** (1905), pp. 143–4, English translation quoted in John Stachel (ed.), *Einstein's Miraculous Year: Five Papers that Changed the Face of Physics*, Princeton University Press, 2005, p. 191. The italics are mine.

5. The Rydberg formula can be written $1/\lambda = R_H(1/m^2 - 1/n^2)$, where λ is the wavelength of the emitted radiation, measured in a vacuum, and R_H is the Rydberg constant for hydrogen.

6. Bohr imposed the condition that the angular momentum of the electron in an orbit around the nucleus be constrained to $nh/2\pi$, where n is an integer number (a quantum number) and h is Planck's constant.

7. We saw in Chapter 5 that in Einstein's derivation of $E = mc^2$ he deduced that the energy of a system measured in its rest frame (call this E_0) increases to the relativistic energy $E = \gamma E_0$ when measured in a frame of reference moving with velocity v. We can re-arrange this expression to give $E/\gamma = E_0$, or $E\sqrt{(1 - v^2/c^2)} = E_0$. Squaring both sides of this equation then gives: $E^2(1 - v^2/c^2) = E_0^2$. If we now multiply through the term in brackets and re-arrange, we get: $E^2 = (E/c)^2v^2 + E_0^2$. Of course, $(E/c)^2 = m^2c^2$, so $E^2 = (mv)^2c^2 + E_0^2$. For speeds v much less than c the product mv is the momentum of the object, given the symbol p. We generalize p to represent the relativistic momentum. Finally, we have $E^2 = p^2c^2 + E_0^2$. We can now substitute for $E_0 = m_0c^2$ to give $E^2 = p^2c^2 + m_0^2c^4$. This is the full expression for the relativistic energy of radiation and matter. If this expression is unfamiliar and rather daunting, think of it this way. If the relativistic energy E is the hypotenuse of a right-angled triangle, then the kinetic energy pc and the rest energy m_0c^2 form the other two sides. The expression is then just a statement of Pythagoras' theorem. For photons with $m_0 = 0$, this general equation reduces to $E = pc$.

8. You might wonder why the kinetic energy is equal to pc and not $\frac{1}{2}pc$, which would appear to be more consistent with the classical expression

for kinetic energy, $\frac{1}{2}mv^2$ (which we can re-write as $\frac{1}{2}pv$, where $p = mv$). Are we missing a factor of $\frac{1}{2}$? Actually, no. If we're prepared to make a few assumptions, we can derive a simpler expression for the relativistic energy for situations where the speed v is much less than c. Recall from Chapter 5 (n 7) that for speeds v much less than c we can approximate γ as $1 + \frac{1}{2}v^2/c^2$. Using this simplified version in $E = \gamma E_0$ gives $E = E_0 + \frac{1}{2}E_0 v^2/c^2$. Again, we can substitute $E_0 = m_0 c^2$ to give $E = m_0 c^2 + \frac{1}{2}m_0 v^2$, which is the rest energy plus the classical kinetic energy with speed v. Setting the linear momentum $p = mv$ gives us $E = m_0 c^2 + \frac{1}{2}pv$.

9. De Broglie wrote: 'After long reflection in solitude and meditation, I suddenly had the idea, during the year 1923, that the discovery made by Einstein in 1905 should be generalized by extending it to all material particles and notably to electrons.' Louis de Broglie, from the 1963 re-edited version of his Ph.D. thesis, quoted in Abraham Pais, *Subtle is the Lord: The Science and the Life of Albert Einstein*, Oxford University Press, 1982, p. 436.

10. In 2012, the Australian tennis player Sam Groth served an ace with a recorded speed of 263.4 kilometres per hour. We can covert this to 73 metres per second. The mass of a tennis ball is typically in the range 57–59 grams, so let's run with 58 grams, or 0.058 kilograms. This gives a linear momentum $p = mv$ for the ball in flight of 0.058 × 73 = 4.2 kgms^{-1}. We can now use $\lambda = h/p$ to calculate the wavelength of the tennis ball. We get: $\lambda = 6.63 \times 10^{-34}/4.2$ metres, or 1.6×10^{-34} metres. Compare this with wavelengths typical of X-rays, which range between 0.01 and 10×10^{-9} metres. The wavelength of the tennis ball is much, much shorter than the wavelengths of X-rays.

CHAPTER 10: THE WAVE EQUATION

1. Felix Bloch, *Physics Today*, **29**, 1976, p. 23.
2. In modern atomic physics, the azimuthal quantum number l takes values $l = 0, 1, 2$, and so on up to the value of $n - 1$. So, if $n = 1$ then l can take only one value, $l = 0$. This corresponds to an electron orbital labelled 1s (the label 's' is a hangover from the early days of atomic spectroscopy when the absorption or emission lines were labelled 's' for 'sharp', 'p' for 'principal', 'd' for 'diffuse', and so on).

For $n = 2$, l can take the values 0 (corresponding to the 2s orbital) and 1 (2p). The magnetic quantum number m takes integer values in the series $-l, \ldots 0, \ldots, +l$. So, if $l = 0$, $m = 0$ and there is only one s orbital, irrespective of the value of n. But when $l = 1$ there are three possible orbitals corresponding to $m = -1$, $m = 0$, and $m = +1$. There are therefore three p orbitals. In the case of the 2p orbitals these are sometimes shown mapped to Cartesian co-ordinates as $2p_x$, $2p_y$ and $2p_z$.

3. Bloch, *Physics Today* (n 1), p. 23.

4. A sine wave moving to the right in the positive x-direction has a general form $\sin(kx - \omega t)$, where k is the 'wave vector' given by $2\pi/\lambda$ and ω is the 'angular frequency' given by $2\pi\nu$, where λ and ν are the wavelength and frequency of the wave. It might not be immediately obvious that this represents a wave moving to the right, in the positive x-direction, so let's take a closer look at it. Let's call the location of the first peak of this wave as measured from the origin x_{peak}. At this point $\sin(kx_{peak} - \omega t) = 1$, or alternatively the angle given by $kx_{peak} - \omega t$ is equal to $90°$ (or $(\frac{1}{2})\pi$ radians). We know that $k = 2\pi/\lambda$ and $\omega = 2\pi\nu$, so we can rearrange this expression to give $x_{peak} = \nu\lambda t + \lambda/4$. Of course, ν times λ is just the wave velocity, v. At a time $t = 0$, the first peak of the wave appears at a distance $x_{peak} = \lambda/4$: the wave rises to its peak in the first quarter of its cycle, before falling again, as in $\sim$. As time increases from zero, we see that x_{peak} *increases* by a distance given by vt. In other words, the wave moves to the right.

5. Erwin Schrödinger, *Annalen der Physik*, **79** (1926), p. 361. Quoted in Walter Moore, *Schrödinger: Life and Thought*, Cambridge University Press, 1989, p. 202.

6. We can rewrite $\frac{1}{2}mv^2$ in terms of the classical linear momentum, $p = mv$, as $\frac{1}{2}p^2/m$. Schrödinger's wave equation could be interpreted to mean that p^2 in the kinetic energy term had been replaced by a differential operator.

7. Let's just prove that to ourselves. We'll use the function x^2. The two mathematical operations we'll perform are 'multiply by 2' and 'take the square root'. If we multiply by 2 first, then the result is simply $\sqrt{2x^2}$ or $1.414x$. If we take the square root first and then multiply by 2, we get $2x$.

8. Heisenberg wrote: 'I remember discussions with Bohr which went through many hours till very late at night and ended almost in despair;

and when at the end of the discussion I went alone for a walk in the neighbouring park I repeated to myself again and again the question: Can nature possibly be as absurd as it seemed...?' Werner Heisenberg, *Physics and Philosophy: The Revolution in Modern Science*, Penguin, London, 1989 (first published 1958), p. 30.

9. The frequency of the wave is given by its speed divided by its wavelength, $\nu = v/\lambda$, where v is the velocity. Alternatively, $\lambda = v/\nu$, which we can substitute into the de Broglie relationship $\lambda = h/p$ to give $v/\nu = h/p$. Rearranging, we get $\nu = pv/h$.

10. The 1s orbital has $n = 1$ and $l = 0$ and possesses a spherical shape. This can accommodate up to two electrons (accounting for hydrogen and helium). For $n = 2$ we have a spherical 2s ($l = 0$) and three dumbbell-shaped 2p ($l = 1$) orbitals, accommodating up to a total of eight electrons (lithium to neon). For $n = 3$ we have one 3s ($l = 0$), three 3p ($l = 1$), and five 3d ($l = 2$) orbitals. These can accommodate up to eighteen electrons but it turns out that the 4s orbital actually lies somewhat lower in energy than 3d and is filled first. The pattern is therefore 3s and 3p (eight electrons—sodium to argon), then 4s, 3d, and 4p (eighteen electrons—potassium to krypton).

CHAPTER 11: THE ONLY MYSTERY

1. Richard P. Feynman, Robert B. Leighton, and Matthew Sands, *The Feynman Lectures on Physics*, Vol. III, Addison-Wesley, Reading, Massachusetts, 1965, p. 1–1.

2. Paul Dirac, *Proceedings of the Royal Society*, **A133**, 1931, pp. 60–72, quoted in Helge S. Kragh, *Dirac: A Scientific Biography*, Cambridge University Press, 1990, p. 103.

3. The catalogue can be found at http://pdg.lbl.gov/. Select 'Summary Tables' from the menu and then 'Leptons'. The electron tops this list.

4. Einstein wrote: 'Quantum mechanics is very impressive. But an inner voice tells me that it is not yet the real thing. The theory produces a good deal but hardly brings us closer to the secret of the Old One. I am at all events convinced that *He* does not play dice.' Albert Einstein, letter to Max Born, 4 December 1926, quoted in Abraham Pais, *Subtle is the Lord: The Science and the Life of Albert Einstein*, Oxford University Press, 1982, p. 443.

5. Albert Einstein, Boris Podolsky, and Nathan Rosen, *Physical Review*, **47**, 1935, pp. 777–80. This paper is reproduced in John Archibald Wheeler and Wojciech Hubert Zurek, (eds), *Quantum Theory and Measurement*, Princeton University Press, 1983, p. 141.

6. Bell wrote: 'If the [hidden variable] extension is local it will not agree with quantum mechanics, and if it agrees with quantum mechanics it will not be local. This is what the theorem says.' John Bell, *Epistemological Letters*, November 1975, pp. 2–6. This paper is reproduced in J.S. Bell, *Speakable and Unspeakable in Quantum Mechanics*, Cambridge University Press, 1987, pp. 63–6. The quote appears on p. 65.

7. E.g., for one specific experimental arrangement, the generalized form of Bell's inequality demands a value that cannot be greater than 2. Quantum theory predicts a value of $2\sqrt{2}$, or 2.828. The physicists obtained the result 2.697 ± 0.015. In other words, the experimental result exceeded the maximum limit predicted by Bell's inequality by almost fifty times the experimental error, a powerful, statistically significant violation.

8. A.J. Leggett, *Foundations of Physics*, **33**, 2003, pp. 1474–5.

9. For a specific experimental arrangement, the whole class of crypto non-local hidden variable theories predicts a maximum value for the Leggett inequality of 3.779. Quantum theory violates this inequality, predicting a value of 3.879, a difference of less than three per cent. The experimental result was 3.852 ± 0.023, a violation of the Leggett inequality by more than three times the experimental error.

10. In these experiments for a specific arrangement the maximum value allowed by the Leggett inequality is 1.78868, compared with the quantum theory prediction of 1.93185. The experimental result was 1.9323 ± 0.0239, a violation of the inequality by more than six times the experimental error.

CHAPTER 12: MASS BARE AND DRESSED

1. Hans Bethe, Calculating the Lamb Shift, *Web of Stories*, http://www. webofstories.com/play/hans.bethe/104;jsessionid=45C0C719DE8C EA2C0899D6A63E281F24.

2. Although structurally very different, we can get some idea of how the perturbation series is supposed to work by looking at the power series expansion for a simple trigonometric function such as sin x. The first few terms in the expansion are: sin $x = x - x^3/3! + x^5/5! - x^7/7! + \ldots$. In this equation, 3! means 3-factorial, or $3 \times 2 \times 1$ (= 6); $5! = 5 \times 4 \times 3 \times 2 \times 1$ (= 120), and so on. For $x = 45°$ (0.785398 radians), the first term (x) gives 0.785398, from which we subtract $x^3/3!$ (0.080745), then add $x^5/5!$ (0.002490), then subtract $x^7/7!$ (0.000037). Each successive term gives a smaller correction, and after just four terms we have the result 0.707106, which should be compared with the correct value, sin(45°) = 0.707107.

3. Murray Gell-Mann, *Nuovo Cimento*, Supplement No. 2, Vol. 4, Series X, 1958, pp. 848–66. In a footnote on p. 859, he wrote: '…any process not forbidden by a conservation law actually does take place with appreciable probability. We have made liberal and tacit use of this assumption, which is related to the state of affairs that is said to prevail in a perfect totalitarian state. Anything that is not compulsory is forbidden.' He was paraphrasing Terence Hanbury (T.H.) White, the author of *The Once and Future King*.

CHAPTER 13: THE SYMMETRIES OF NATURE

1. Wolfgang Pauli, part of a conversation reported by Chen-Ning Yang at the International Symposium on the History of Particle Physics, Batavia, Illinois, 2 May 1985, quoted by Michael Riordan, *The Hunting of the Quark: A True Story of Modern Physics*, Simon & Shuster, New York, 1987, p. 198.

2. It is relatively straightforward to picture the symmetry transformations of U(1) in the so-called complex plane, the two-dimensional plane formed by one real axis and one 'imaginary axis'. The imaginary axis is constructed from real numbers multiplied by i, the square root of minus one. We can pinpoint any complex number z in this plane using the formula, $z = re^{i\theta}$, where r is the length of the line joining the origin with the point z in the plane and θ is the angle between this line and the real axis. This expression for z can be re-written using Euler's formula as $z = r(\cos\theta + i\sin\theta)$, which makes the connection between U(1) and continuous transformations in a circle and with the phase angle of a sine wave.

3. We can get a very crude sense for why this must be from Heisenberg's energy–time uncertainty relation and special relativity. According to the uncertainty principle the product of the uncertainties in energy and the rate of change of energy with time, $\Delta E\Delta t$, cannot be less than $h/4\pi$. The range of a force-carrying particle is then roughly determined by the distance it can travel in the time Δt. We know that nothing can travel faster than the speed of light so the *maximum* range of a force-carrying particle is given by $c\Delta t$, or $hc/4\pi\Delta E$. If we approximate ΔE as mc^2, where m is the mass of the force carrier, then the range (let's call it R) is given by $h/4\pi mc$. We see that the range of the force is inversely proportional to the mass of the force carrier. If we assume photons are massless ($m = 0$), then the range of the electromagnetic force is infinite.

4. The radius of a proton is something of the order of 0.85×10^{-15} m. If we assume that the force binding protons and neutrons together inside the nucleus must operate over this kind of range, we can crudely estimate the mass of the force-carrying particles that would be required from $h/4\pi Rc$. Plugging in the values of the physical constants gives us a mass of 0.2×10^{-27} kg, or ~100 MeV/c^2, about eleven per cent of the mass of a proton. This is the figure obtained by Yukawa in 1935 for particles that he believed should carry the strong force between protons and neutrons. Although the strong force is now understood to operate very differently (see Chapter 15), Yukawa was correct in principle. These 'force carriers' are the *pions*, which come in positive (π^+), negative (π^-) and neutral (π^0) varieties with masses of about 140 MeV/c^2 (π^+ and π^-) and 135 MeV/c^2 (π^0).

5. Riordan, *The Hunting of the Quark* (n 1), p. 198.

6. Yang wrote: 'The idea was *beautiful* and should be published. But what is the mass of the [force-carrying] gauge particle? We did not have firm conclusions, only frustrating experiences to show that [this] case is much more involved than electromagnetism. We tended to believe, on physical grounds, that the charged gauge particles cannot be massless.' Chen Ning Yang, *Selected Papers with Commentary*, W.H. Freeman, New York, 1983, quoted by Christine Sutton in Graham Farmelo, (ed.), *It Must be Beautiful: Great Equations of Modern Science*, Granta Books, London, 2002, p. 243.

7. I've scratched around in an attempt to find a simple and straightforward explanation for what is a very important feature of quantum field theories, but have come to the conclusion that this is really difficult to do without resorting to a short course on the subject. The best I can do is give you a sense for where the 'mass term' comes from. Remember that Schrödinger derived his non-relativistic wave equation from the equation for classical wave motion by substituting for the wavelength using the de Broglie relation, $\lambda = h/p$. His manipulations had the effect of changing the nature of the kinetic energy term in the equation for the total energy. In Newtonian mechanics this is the familiar $\frac{1}{2}mv^2$, where m is the mass and v the velocity. We can re-write this in terms of the linear momentum p ($= mv$) as $\frac{1}{2}p^2/m$. In the Schrödinger wave equation, the expression for kinetic energy is structurally similar but the classical p^2 is now replaced by a mathematical operator (let's call it $\mathbf{p}^2$, which means the operator is applied twice to the wavefunction, ψ). But, as we saw in Chapter 10, Schrödinger's equation does not conform to the demands of special relativity. In fact, Schrödinger did work out a fully relativistic wave equation but found that it did not make predictions that agreed with experiment. This version of the wave equation was rediscovered by Swedish theorist Oskar Klein and German theorist Walter Gordon in 1926 and is known as the Klein–Gordon equation. Its derivation is based on the expression for the relativistic energy, $E^2 = p^2c^2 + m_0^2c^4$, which is 'quantized' by replacing the classical p^2 with the quantum-mechanical operator equivalent, $\mathbf{p}^2$, just as Schrödinger had done. There are two things to note about this. First, we're dealing not with energy but with energy-squared. Second, we have now introduced a term which depends on the square of the mass. We introduce a quantum field, ϕ, on which the momentum operator is applied, and in consequence a term in the dynamical equations appears which is related to $m^2\phi^2$. Although the Klein–Gordon equation does not account for spin (so it can't be used to describe electrons, as Schrödinger discovered), it is perfectly valid when applied to particles with zero spin (as it happens, particles such as the pions). From it we learn that mass terms related to $m^2\phi^2$ can be expected to appear in any valid formulation of quantum field theory that meets the demands of special relativity.

CHAPTER 14: THE GODDAMN PARTICLE

1. Leon Lederman, with Dick Teresi, *The God Particle: If the Universe is the Answer, What is the Question?*, Bantam Press, London, 1993, p. 22.

2. Look back at Chapter 12, n 3. We can crudely estimate the range of a force carried by a particle with a mass this size from $h/4\pi mc$. Let's set $m = 350 \times 10^{-27}$ kg (a couple of hundred times the proton mass) and plug in the values of the physical constants. We get a range of about 0.5×10^{-18} m, well *inside* the confines of the proton or neutron (which, remember, have a radius of about 0.85×10^{-15} m). A recent paper by the ZEUS Collaboration at the Hadron-Elektron Ring Anlage (HERA) in Hamburg, Germany recently set an upper limit on the size of a quark of 0.43×10^{-18} m. This article is available online at the Cornell University Library electronic archive site (http://arxiv.org), with the reference arXiv: 1604.01280v1, 5 April 2016. Actually, it turns out that the weak force carriers are not quite this heavy (they are a little less than 100 times the mass of a proton).

3. Schwinger later explained: 'It was numerology...But—that's the whole idea. I mentioned this to [J. Robert] Oppenheimer, and he took it very coldly, because, after all, it was an outrageous speculation.' Comment by Julian Schwinger at an interview on 4 March 1983, quoted in Robert P. Crease and Charles C. Mann, *The Second Creation: Makers of the Revolution in Twentieth-century Physics*, Rutgers University Press, 1986, p. 216.

4. Some years later, Nambu wrote: 'What would happen if a kind of superconducting material occupied all of the universe, and we were living in it? Since we cannot observe the true vacuum, the [lowest-energy] ground state of this medium would become the vacuum, in fact. Then even particles which were massless...in the true vacuum would acquire mass in the real world.' Yoichiro Nambu, *Quarks*, World Scientific Publishing, Singapore, 1981, p. 180.

5. Higgs wrote: 'I was indignant. I believed that what I had shown could have important consequences in particle physics. Later, my colleague Squires, who spent the month of August 1964 at CERN, told me that the theorists there did not see the point of what I had done. In retrospect, this is not surprising: in 1964...quantum field theory was out of fashion...'. Peter Higgs, in Lillian Hoddeson, Laurie Brown, Michael

Riordan, and Max Dresden, *The Rise of the Standard Model: Particle Physics in the 1960s and 1970s*, Cambridge University Press, 1997, p. 508.

6. In his Nobel lecture, Weinberg said: 'At some point in the fall of 1967, I think while driving to my office at MIT, it occurred to me that I had been applying the right ideas to the wrong problem.' Steven Weinberg, *Nobel Lectures, Physics 1971–1980*, ed. Stig Lundqvist, World Scientific, Singapore (1992), p. 548.

7. In an interview with Robert Crease and Charles Mann on 7 May 1985, Weinberg declared: 'My God, this is the answer to the weak interaction!' Steven Weinberg, quoted in Crease and Mann, *The Second Creation* (n 3), p. 245.

8. In his Foreword to my book *Higgs*, Weinberg wrote: 'Rather, I did not include quarks in the theory simply because in 1967 I just did not believe in quarks. No-one had ever observed a quark, and it was hard to believe that this was because quarks are much heavier than observed particles like protons and neutrons, when these observed particles were supposed to be made of quarks.' See Jim Baggott, *Higgs: The Invention and Discovery of the 'God' Particle*, Oxford University Press, 2012, p. xx.

9. Lederman, with Teresi, *The God Particle* (n 1), p. 22.

CHAPTER 15: THE STANDARD MODEL

1. Willis Lamb, *Nobel Lectures, Physics 1942–1962*, Elsevier, Amsterdam (1964), p. 286.

2. Enrico Fermi, quoted as 'physics folklore' by Helge Kragh, *Quantum Generations: A History of Physics in the Twentieth Century*, Princeton University Press, 1999, p. 321.

3. Murray Gell-Mann and Edward Rosenbaum, *Scientific American*, July 1957, pp. 72–88.

4. Murray Gell-Mann, interview with Robert Crease and Charles Mann, 3 March 1983, quoted in Robert P. Crease and Charles C. Mann, *The Second Creation: Makers of the Revolution in Twentieth-century Physics*, Rutgers University Press, 1986, p. 281.

5. Gell-Mann said: 'That's it! Three quarks make a neutron and a proton!', interview with Robert Crease and Charles Mann, 3 March 1983, quoted in Crease and Mann, ibid., p. 282.

6. 'We gradually saw that that [colour] variable was going to do every-thing for us!', Gell-Mann explained. 'It fixed the statistics, and it could do that without involving us in crazy new particles. Then we realized that it could also fix the dynamics, because we could build an SU(3) gauge theory, a Yang-Mills theory, on it.' W.A. Bardeen, H. Fritzsch, and M. Gell-Mann, *Proceedings of the Topical Meeting on Conformal Invariance in Hadron Physics*, Frascati, May 1972, quoted in Crease and Mann, *The Second Creation* (n 4), p. 328.

7. Joe Incandela, CERN Press Release, 14 March 2013.

8. 'The data are consistent with the Standard Model predictions for all parameterisations considered.' ATLASCONF-2015-044/CMS-PAS-HIG-15-002, 15 September 2015.

CHAPTER 16: MASS WITHOUT MASS

1. Frank Wilczek, 'Four Big Questions with Pretty Good Answers', talk given at a Symposium in Honour of Heisenberg's 100th birthday, 6 December 2001, Munich. This article is available online at the Cornell University Library electronic archive site (http://arxiv.org), with the reference arXiv:hep-ph/0201222v2, 5 February 2002. See also 'QCD and Natural Philosophy', arXiv:physics/0212025v2 [physics.ed-ph], 12 December 2002.

2. Lucretius, *On the Nature of the Universe*, trans. R.E. Latham, Penguin Books, London, first published 1951, p. 189.

3. Lucretius wrote: 'The more the earth is drained of heat, the colder grows the water, the colder grows the water embedded in it.' Lucretius, ibid., p. 243.

4. Each side measures 2.7 centimetres, so the volume of the cube of ice is $2.7^3 = 19.7$ cubic centimetres. If we look up the density of pure ice at 0°C we find this is 0.9167 grams per cubic centimetre. So, the mass (which I'll not distinguish from weight) of the cube of ice is given by the density multiplied by the volume, or $0.9167 \times 19.7 = 18.06$ grams.

5. S. Durr, Z. Fodor, J. Frison, *et al.*, *Science*, **322**, pp. 1224–1227, 21 November 2008, arXiv:0906.3599v1 [hep-lat], 19 June 2009.

6. Sz. Borsanyi, S. Durr, Z. Fodor, *et al.*, *Science*, **347**, pp. 1452–1455, 27 March 2015, also available as arXiv:1406.4088v2 [hep-lat], 7 April 2015.

See also the commentary by Frank Wilczek, *Nature*, **520**, pp. 303–4, 16 April 2015.

7. Wilczek, 'Four Big Questions' (n 1).

8. See Frank Wilczek, *MIT Physics Annual 2003*, MIT, pp. 24–35.

9. Albert Einstein, *Annalen der Physik*, **18** (1905), pp. 639–41, trans. and repr. in John Stachel (ed.), *Einstein's Miraculous Year: Five Papers that Changed the Face of Physics*, centenary edn, Princeton University Press, 2005. The quote appears on p. 164.

10. Frank Wilczek, *The Lightness of Being*, Penguin, London, 2008, p. 132.

EPILOGUE

1. Albert Einstein and Leopold Infeld, *The Evolution of Physics*, Simon & Schuster, New York, 1938, p. 313, quoted in Alice Calaprice, *The Ultimate Quotable Einstein*, Princeton University Press, 2011, p. 390.

2. Max Jammer, *Concepts of Mass in Contemporary Physics and Philosophy*, Princeton University Press, 2000, p. 167.

3. Stephen P. Martin, 'A Supersymmetry Primer', version 6, arXiv: hep-ph/9709356, September 2011, p. 5.

4. The Planck length is given by $\sqrt{(hG/2\pi c^3)}$, where h is Planck's constant, G is Newton's gravitational constant, and c is the speed of light.

5. Lee Smolin, *Three Roads to Quantum Gravity: A New Understanding of Space, Time and the Universe*, Phoenix, London, 2001, p. 211.

6. Carlo Rovelli, 'Loop Quantum Gravity: The First Twenty-five Years', arXiv: 1012.4707v5 [gr-qc], 28 January 2012, p. 20.

7. Carlo Rovelli, ibid., p. 7.

8. In *Reality and the Physicist*, the philosopher Bernard d'Espagnat wrote: '...we must conclude that physical realism is an "ideal" from which we remain distant. Indeed, a comparison with conditions that ruled in the past suggests that we are a great deal more distant from it than our predecessors thought they were a century ago', *Reality and the Physicist: Knowledge, Duration and the Quantum World*, Cambridge University Press, 1989, p. 115.

9. In 1900, the great British physicist Lord Kelvin (William Thomson) is supposed to have famously declared to the British Association for

the Advancement of Science that: 'There is nothing new to be discovered in physics now. All that remains is more and more precise measurement.' Whilst it appears that we have no evidence that Kelvin ever said this, in *Light Waves and their Uses*, published in 1903 by The University of Chicago Press, American physicist Albert Michelson wrote:

> Many other instances might be cited, but these will suffice to justify the statement that 'our future discoveries must be looked for in the sixth place of decimals.' It follows that every means which facilitates accuracy in measurement is a possible factor in a future discovery, and this will, I trust, be a sufficient excuse for bringing to your notice the various methods and results which form the subject-matter of these lectures.

This quote appears on pp. 24–5. It is thought that Michelson may have been quoting Kelvin.

GLOSSARY

acceleration The rate of change of velocity (speed) with time, usually given the symbol a, as in $F = ma$, or force equals mass times acceleration (Newton's second law).

anti-particle Identical in mass to an 'ordinary' particle but of opposite charge. For example, the anti-particle of the electron (e^-) is the positron (e^+). The anti-particle of a red quark is an anti-red anti-quark. Every particle in the standard model has an anti-particle. Particles with zero charge are their own anti-particles.

ATLAS Acronym for A Toroidal LHC Apparatus, one of the two main detector collaborations involved in the hunt for the Higgs boson at CERN's Large Hadron Collider.

atom From the Greek *atomos*, meaning 'indivisible' or 'uncuttable'. Originally intended to denote the ultimate constituents of matter, the word 'atom' now signifies the fundamental constituents of individual chemical elements. Thus, water consists of molecules of H_2O, which is composed of two atoms of hydrogen and one atom of oxygen. The atoms in turn consist of protons and neutrons, which are bound together to form a central nucleus, and electrons whose wavefunctions form characteristic patterns called orbitals around the nucleus.

atomism A natural philosophy based on the notion that matter is composed of ultimate, indivisible, uncuttable parts, called atoms. The atomist tradition is most closely associated with the ancient Greek philosophers Leucippus, Democritus, and Epicurus, and the Roman poet and philosopher Lucretius, although atomism also formed part of the philosophy of certain schools in ancient India.

atto A prefix denoting a billion billionth (10^{-18}). An attometre (am) is 10^{-18} metres, or a thousandth of a femtometre. The radius of a proton is about 850 am. The LIGO gravitational wave observatory is sensitive to displacements of the order of 1 am.

Avogadro's constant Typically given the symbol N_A. Defined as the number of atoms in 12 grams (or 1 mole) of carbon-12. It has the value 6.022×10^{23} per mole.

Avogadro's hypothesis/law Equal volumes of all gases, at the same temperature and pressure, contain equal numbers of particles (atoms or molecules). This follows because, with fixed temperature and pressure, the volume of a gas is directly proportional to the *amount* of gas present (e.g., as measured by its weight). One mole of gas at a temperature of 273.15 kelvin (0 °C) and pressure of 101.325 thousand Pascals (1 atmosphere) has a volume of about 22.4 litres, and contains 6.022×10^{23} atoms or molecules.

bare mass The hypothetical mass that a particle would possess if it could be separated from the quantum fields which it generates, or with which it interacts. The observed mass of the particle is then the bare mass plus mass generated by interactions with the quantum fields.

baryon From the Greek *barys*, meaning 'heavy'. Baryons form a sub-class of hadrons. They are heavier particles which experience the strong nuclear force and include the proton and neutron. They are comprised of triplets of quarks.

Bell's theorem/inequality Devised by John Bell in 1966. The simplest extension of quantum theory, which resolves the problem of the collapse of the wavefunction and the 'spooky action-at-a-distance' that this seems to imply, involves the introduction of local hidden variables which govern the properties and behaviour of quantum particles. Bell's theorem states that the predictions of any local hidden variable theory will not always agree with the predictions of quantum theory. This is summarized in Bell's inequality—the predictions of local hidden variable theories cannot exceed a certain maximum limit. But quantum theory predicts results that for certain experimental arrangements *will* exceed this limit. Bell's inequality therefore allows a direct experimental test.

beta particle A high-speed electron emitted from the nucleus of an atom undergoing beta radioactive decay. *See* beta radioactivity/decay.

beta radioactivity/decay First discovered by French physicist Henri Becquerel in 1896 and so named by Ernest Rutherford in 1899. An example of a weak force decay, it involves transformation of a down quark in a neutron into an up quark, turning the neutron into a proton with the emission of a W⁻ particle. The W⁻ decays into a high-speed electron (the 'beta particle') and an electron anti-neutrino.

big bang Term used to describe the cosmic 'explosion' of spacetime and matter during the early moments in the creation of the universe, about 13.8 billion years ago. Originally coined by maverick physicist Fred Hoyle as a derogatory term, overwhelming evidence for a big bang 'origin' of the universe has since been obtained through the detection and mapping of the cosmic microwave background radiation, the cold remnant of hot radiation thought to have disengaged from matter about 380,000 years after the big bang.

billion One thousand million, 10^9, or 1,000,000,000.

black hole A name popularized (though not, as many suggest, coined) by John Wheeler. A black hole is a region of spacetime containing so much mass-energy that its escape velocity—the speed required to escape its gravitation pull—is greater than the speed of light. This idea actually dates back to the eighteenth century, but came to prominence through the work of Karl Schwarzschild, who in 1916 was the first to derive formal solutions for Einstein's gravitational field equations. *See also* Schwarzschild solution/radius.

Bohr radius The orbital distance of the electron as measured from the proton in a hydrogen atom. In Bohr's model of the atom, published in 1913, Bohr calculated this distance from a collection of fundamental physical constants (including Planck's constant, the speed of light, and the mass and charge of the electron). In Schrödinger's wave mechanics, the electron is spherically 'distributed' within its lowest-energy orbital, but has the highest probability for being found at a distance corresponding to the Bohr radius, a little over 0.0529 nanometres.

Boltzmann constant Typically denoted k or k_B. This constant connects the energy (E) of individual particles with temperature (T), such that $T = E/k_B$ (or $E = k_B T$), and was actually first introduced by Planck in his derivation of the radiation law. Even the iconic equation connecting entropy (S) and probability (W) which is carved on Boltzmann's headstone, $S = k\ln W$, was derived by Planck, not Boltzmann.

boson Named for Indian physicist Satyendra Nath Bose. Bosons are characterized by integral spin quantum numbers (1, 2,..., etc.) and, as such, are not subject to Pauli's exclusion principle. Bosons are involved in the transmission of forces between matter particles, and include the photon (electromagnetism), the W and Z particles (weak force), and gluons (colour force). Particles with spin zero are also called bosons but these are not involved in transmitting forces. Examples include the pions and the Higgs boson. The graviton, the hypothetical particle of the gravitational field, is believed to be a boson with spin 2.

bottom quark Also sometimes referred to as the 'beauty' quark. A third-generation quark with charge $-\frac{1}{3}$, spin $\frac{1}{2}$ (fermion), and a mass of 4.18 GeV/c^2. It was discovered at Fermilab in 1977, through the observation of the upsilon, a meson formed from bottom and anti-bottom quarks.

CERN Acronym for Conseil Européen pour la Recherche Nucléaire (the European Council for Nuclear Research), founded in 1954. This was renamed the Organisation Européenne pour la Recherche Nucléaire (European Organization for Nuclear Research) when the provisional Council was dissolved, but the acronym CERN was retained. CERN is located in the north-west suburbs of Geneva near the Swiss–French border.

charm quark A second generation quark with charge $+\frac{2}{3}$, spin $\frac{1}{2}$ (fermion) and a mass of 1.28 GeV/c^2. It was discovered simultaneously at Brookhaven National Laboratory and SLAC in the 'November revolution' of 1974 through the observation of the J/ψ, a meson formed from a charm and anti-charm quark.

classical mechanics The system of mechanics embodied in Newton's laws of motion and the law of universal gravitation, although the study of mechanics predates Newton. The system deals with the influence of forces on the motions of larger, macroscopic bodies at speeds substantially less than light. Although 'classical', the system remains perfectly valid today within the limits of its applicability.

classical modern philosophy The philosophy that emerged in Europe in the seventeenth and eighteenth centuries as the grip of the establish Church began to be relaxed is sometimes referred to as 'classical modern'. It succeeded medieval philosophy. Classical modern philosophers include: René Descartes (born 1596), John Locke (1632), Baruch Spinoza

(1632), Gottfried Leibniz (1646), George Berkeley (1685), David Hume (1711), and Immanuel Kant (1724).

CMS Acronym for Compact Muon Solenoid, one of the two detector collaborations involved in the hunt for the Higgs boson at CERN's Large Hadron Collider.

cold dark matter A key component of the current Λ-CDM model of big bang cosmology, thought to account for about twenty-six per cent of the mass-energy of the universe. The constitution of cold dark matter is unknown, but is thought to consist largely of 'non-baryonic' matter, i.e. matter that does not involve protons or neutrons, most likely particles not currently known to the standard model.

collapse of the wavefunction In most quantum systems, the wavefunction of a quantum entity will be delocalized over a region of space (the quantum entity may be here, there, or everywhere within the boundary of the wavefunction), yet when a measurement is made the result is localized to a specific position (the entity is here). Similarly, in a quantum measurement in which a number of different outcomes is possible (such as spin-up or spin-down), it is necessary to form a superposition of the wavefunctions describing these outcomes. The probability of getting a specific result is then related to the square of the amplitude of the corresponding wavefunction in the superposition. In either case, the wavefunction or superposition is said to 'collapse'. A number of possible outcomes converts to one outcome only, and all the other possibilities disappear.

colour charge A property possessed by quarks in addition to flavour (up, down, strange, etc.). Unlike electric charge, which comes in two varieties—positive and negative—colour charge comes in three varieties, which physicists have chosen to call red, green, and blue. Obviously, the use of these names does not imply that quarks are 'coloured' in the conventional sense. The colour force between quarks is carried by coloured gluons.

colour force The strong force responsible for binding quarks and gluons together inside hadrons. Unlike more familiar forces, such as gravity and electromagnetism, which get weaker with increasing distance, the colour force acts like a piece of elastic or spring, tethering the quarks together. When the quarks are close together, the elastic or spring is relaxed and the quarks behave as though they are entirely free. But as the quarks are pulled apart, the elastic or spring tightens and keeps the quarks 'confined'. The strong nuclear force which binds protons and neutrons together inside atomic nuclei results from a kind of 'leakage' of the colour force beyond the boundaries of the nucleons.

complementarity The principle of complementarity was devised by Niels Bohr and is a key pillar of the Copenhagen interpretation of quantum mechanics. According to this principle, quantum wave-particles will exhibit wave-like and particle-like behaviour in mutually exclusive experimental arrangements, but it is impossible to devise an arrangement that will show both types of behaviour simultaneously. Such behaviour is, however, not contradictory; it is complementary.

complex number A complex number is formed by multiplying a real number by the square root of −1, written i. The square of a complex number is then a negative number: for example, the square of $5i - (5i)^2$—is −25. Complex numbers are used widely in mathematics to solve problems that are impossible using real numbers only.

conservation law A physical law which states that a specific measureable property of an isolated system does not change as the system evolves in time. Measureable properties for which conservation laws have been established include mass-energy, linear and angular momentum, electric and colour charge, isospin, etc. According to Noether's theorem, each conservation law can be traced to a specific continuous symmetry of the system.

Copenhagen interpretation Developed by Niels Bohr, Werner Heisenberg, and Wolfgang Pauli as a way of thinking about the nature of elementary quantum wave-particles as described by quantum mechanics. Depending on the experimental set-up, such wave-particles will exhibit wave-like or particle-like behaviour. But these behaviours are complementary: in this kind of experiment the wave-particle is a wave, in this other kind of experiment it is a particle and it is meaningless to ask what the wave-particle *really is*.

cosmic background radiation Some 380,000 years after the big bang, the universe had expanded and cooled sufficiently to allow hydrogen nuclei (protons) and helium nuclei (consisting of two protons and two neutrons) to recombine with electrons to form neutral hydrogen and helium atoms. At this point, the universe became 'transparent' to the residual hot radiation. Further expansion has shifted and cooled this hot radiation to the microwave and infrared regions with a temperature of just 2.7 kelvin (−270.5°C), a few degrees above absolute zero. This microwave background radiation was predicted by several theorists and was discovered accidentally by Arno Penzias and Robert Wilson in 1964. The COBE, WMAP, and Planck satellites have since studied this radiation in detail.

cosmic inflation A rapid exponential expansion of the universe thought to have occurred between 10^{-36} and 10^{-32} seconds after the big bang. Discovered by American physicist Alan Guth in 1980, inflation helps to explain the large-scale structure of the universe that we observe today.

cosmic rays Streams of high-energy charged particles from outer space which wash constantly over the Earth's upper atmosphere. The use of the term 'ray' harks back to the early days of research on radioactivity, when directed streams of charged particles were referred to as 'rays'. Cosmic rays are derived from a variety of sources, including high-energy processes occurring on the surface of the Sun and other stars, and as-yet unknown processes occurring elsewhere in the universe. The energies of cosmic ray particles are typically between 10 MeV and 10 GeV.

cosmological constant Albert Einstein initially resisted the idea that the universe is dynamic—that it could expand or contract—and fudged his equations to produce static solutions. Concerned that conventional gravity would be expected to overwhelm the matter in the universe and cause it to collapse in on itself, Einstein introduced a 'cosmological constant'—a kind of negative or repulsive form of gravity—to counteract the effect. When evidence accumulated that the universe is actually expanding, Einstein regretted his action, calling it the biggest blunder he had ever made in his life. But, in fact, further discoveries in 1998 suggested that the expansion of the universe is actually accelerating. When combined with satellite measurements of the cosmic microwave background radiation these results have led to the suggestion that the universe is pervaded by 'dark energy', accounting for about 69.1 per cent of the mass-energy of the universe. One form of dark energy requires the reintroduction of Einstein's cosmological constant.

dark matter Discovered in 1934 by Swiss astronomer Fritz Zwicky as an anomaly in the measured mass of galaxies in the Coma cluster (located in the constellation Coma Berenices). Zwicky observed that the rotation speeds of galaxies near the cluster edge are much faster than predicted from the mass of all the observable galaxies, implying that the mass of the cluster is actually much larger. As much as ninety per cent of the mass required to explain the rotation curve appeared to be 'missing', or invisible. This missing matter was called 'dark matter'. Subsequent studies favour a form of dark matter called 'cold dark matter'. *See* cold dark matter.

de Broglie relation Deduced by Louis de Broglie in 1923, this equation relates a wave-like property (the wavelength, λ) of a quantum wave-particle to a particle-like property (linear momentum, p). The relation is $\lambda = h/p$, where h is Planck's constant. For 'everyday' objects of macroscopic size (such as a tennis ball), the wavelength predicted by the de Broglie relation is much too short to observe. But microscopic entities such as electrons have measureable wavelengths, typically 100,000 times shorter than visible light. A beam of electrons can be diffracted and will show two-slit interference effects. Electron microscopes are used routinely to study the structures of inorganic and biological samples.

dressed mass The mass derived from a quantum wave-particle's self-energy, the result of interactions with the system from which it is physically inseparable. For example, an electron acquires self-energy by interacting with its own self-generated electromagnetic field.

electric charge A property possessed by quarks and leptons (and, more familiarly, protons and electrons). Electric charge comes in two varieties—positive and negative—and the flow of electrical charge is the basis for electricity and the power industry.

electromagnetic force Electricity and magnetism were recognized to be components of a single, fundamental force through the work of several experimental and theoretical physicists, most notably English physicist Michael Faraday and Scottish theoretician James Clerk Maxwell. The electromagnetic force is responsible for binding electrons with their nuclei inside atoms, and binding atoms together to form the great variety of molecular substances.

electron Discovered in 1897 by English physicist Joseph John Thompson. The electron is a first-generation lepton with a charge –1, spin ½ (fermion), and mass 0.51 MeV/c^2.

electron volt (eV) An electron volt is the amount of energy a single negatively charged electron gains when accelerated through a one-volt electric field. A 100W light bulb burns energy at the rate of about 600 billion billion electron volts per second.

electro-weak force Despite the great difference in scale between the electromagnetic and weak nuclear forces, these are facets of what was once a unified electro-weak force, thought to prevail during the 'electro-weak epoch', between 10^{-36} and 10^{-12} seconds after the big bang. The combination of electromagnetic and weak nuclear forces in a unified field theory was first achieved by Steven Weinberg and independently by Abdus Salam in 1967–1968.

element The philosophers of Ancient Greece believed that all material substance was composed of four elements—earth, air, fire, and water. A fifth element, variously called the ether or 'quintessence', was introduced by Aristotle to describe the unchanging heavens. Today, these classical elements have been replaced by a system of chemical elements. These are 'fundamental' in the sense that chemical elements cannot be transformed one into another by chemical means, meaning that they consist of only one type of atom. The elements are organized in a 'periodic table', from hydrogen to uranium and beyond.

empiricism One of several philosophical perspectives on the acquisition of human knowledge. In an empiricist philosophy, knowledge is firmly linked to experience and evidence—'seeing is believing'. If we can't directly experience or acquire even indirect evidence for the existence of an entity, then we have no grounds for believing that it really exists. Such an entity would then be regarded as metaphysical.

Euclidean space Named for the ancient Greek mathematician Euclid of Alexandria. This is the familiar geometry of 'ordinary' three-dimensional space, typically described in terms of Cartesian co-ordinates (x,y,z) and in which the angles of a triangle add up to 180°, the circumference of a circle is given by 2π times its radius, and parallel lines never meet.

exclusion principle *See* Pauli exclusion principle.

femto A prefix denoting a million billionth (10^{-15}). A femtometre (fm) is 10^{-15} metres, a thousand attometres or a thousandth of a picometre. The radius of a proton is about 0.85 fm.

fermion Named for Italian physicist Enrico Fermi. Fermions are characterized by half-integral spins (½, 3⁄2, etc.) and include quarks and leptons and many composite particles produced from various combinations of quarks, such as baryons.

flavour A property which distinguishes one type of quark from another in addition to colour charge. There are six flavours of quark which form three generations: up, charm and top with electric charge $+\frac{2}{3}$, spin ½ and masses of 1.8–3.0 MeV/c^2, 1.28 GeV/c^2 and 173 GeV/c^2, respectively, and down, strange and bottom with electric charge $-\frac{1}{3}$, spin ½ and masses 4.5–5.3 MeV/c^2, 95 MeV/c^2 and 4.18 GeV/c^2, respectively. The term 'flavour' is also sometimes applied to leptons, with the electron, muon, tau, and their corresponding neutrinos distinguished by their 'lepton flavour'. *See* lepton.

force Any action that changes the motion of an object. In Isaac Newton's three laws of motion forces are 'impressed', meaning that the actions involve some kind of physical contact between the object and whatever is generating the force (such as another object). The exception to the rule is Newton's force of gravity, which appears to act instantaneously and at a distance (no obvious contact between gravitating objects, such as the Earth and the Moon). This problem is resolved in Einstein's general theory of relativity.

g-factor A constant of proportionality between the (quantized) angular momentum of an elementary or composite particle and its magnetic moment, the direction the particle will adopt in a magnetic field. There are actually three g-factors for the electron, one associated with its spin, one associated with the angular momentum of the electron orbital motion in an atom, and one associated with the sum of spin and orbital angular momentum. Dirac's relativistic quantum theory of the electron predicted a g-factor for electron spin of 2. The value recommended in 2010 by the International Council for Science Committee on Data for Science and Technology (CODATA) task group is 2.00231930436153. The difference is due to quantum electrodynamic effects.

general relativity Developed by Einstein in 1915. The general theory of relativity incorporates special relativity and Newton's law of universal gravitation in a geometric theory of gravitation. Einstein replaced the 'action-at-a-distance' implied in Newton's theory with the movement of massive bodies in a curved spacetime. In general relativity, matter tells spacetime how to curve, and the curved spacetime tells matter how to move.

giga A prefix denoting billion. A giga electron volt (GeV) is a billion electron volts, 10^9 eV or 1,000 MeV.

gluon The carrier of the strong colour force between quarks. Quantum chromodynamics requires eight, massless colour force gluons, which themselves carry colour charge. Consequently, the gluons participate in the force rather than simply transmit it from one particle to another. Ninety-nine per cent of the mass of protons and neutrons is thought to be energy carried by gluons and quark–anti-quark pairs created by the colour field.

gravitational force The force of attraction experienced between all mass-energy. Gravity is extremely weak, and has no part to play in the interactions between atoms, sub-atomic, and elementary particles, which are rather governed by the colour force, weak force, and electromagnetism. The effects of gravity are described by Einstein's general theory of relativity and approximated in Newton's theory of universal gravitation.

graviton A hypothetical particle which carries the gravitational force in speculative quantum field theories of gravity. Although many attempts have been made to develop such a theory, to date these have not been recognized as successful. If it exists, the graviton would be a massless, chargeless boson with a spin quantum number of 2.

hadron From the Greek *hadros*, meaning 'thick' or 'heavy'. Hadrons form a class of particles which experience the strong nuclear force and are therefore comprised of various combinations of quarks. This class includes baryons, which are composed of three quarks, and mesons, which are composed of one quark and an anti-quark.

hidden variables The simplest way to modify or extend conventional quantum mechanics to eliminate the collapse of the wavefunction is to introduce hidden variables. Such variables govern the properties and behaviour of quantum wave-particles but by definition cannot be observed directly. If the resulting extension is required to ensure that individual quantum entities possess specific properties at all times (in other words, the entities are 'locally real'), then the hidden variables are said to be local. If the extension is required to ensure that quantum entities possess specific properties in a collective sense, then the hidden variables may be non-local.

Higgs boson Named for English physicist Peter Higgs. All Higgs fields have characteristic field particles called Higgs bosons. The term 'Higgs boson' is typically reserved for the electro-weak Higgs, the particle of the Higgs field first used in 1967–1968 by Steven Weinberg and Abdus Salam to account for electro-weak symmetry-breaking. It is now thought that the electro-weak Higgs boson was discovered at CERN's Large Hadron Collider, a discovery announced on 4 July 2012. It is a neutral, spin-0 particle with a mass of about 125 GeV/c^2.

Higgs field Named for English physicist Peter Higgs. A generic term used for any background quantum field added to a field theory to trigger symmetry-breaking through the Higgs mechanism. The existence of the Higgs field used to break the symmetry in a quantum field theory of the electro-weak force is strongly supported by the discovery of the Higgs boson at CERN.

Higgs mechanism Named for English physicist Peter Higgs, but also often referred to using the names of other physicists who independently discovered the mechanism in 1964: Robert Brout, François Englert, Higgs, Gerald Guralnik, Carl Hagen, and Tom Kibble. The mechanism describes how a background quantum field—called the Higgs field—can be added to a field theory to break a symmetry. In 1967–1968 Steven Weinberg and Abdus Salam independently used the mechanism to develop a field theory of the electro-weak force.

Hubble's law The observation, first reported by American astronomer Edwin Hubble, that distant galaxies recede from us at speeds that are directly proportional to their distances. This relationship is summarized by the equation $v = H_0 D$, in which v is the recession speed of the galaxy, D is its distance from Earth and H_0 is the Hubble constant, which has a value of 67.7 kilometres per second per megaparsec (based on analysis of Planck satellite mission data reported in 2015).

inflation *See* cosmic inflation.

isospin Also known as isotopic or isobaric spin. Introduced by Werner Heisenberg in 1932 to explain the symmetry between the newly discovered neutron and the proton. Isospin symmetry is now understood to be a subset of the more general flavour symmetry in hadron interactions. The isospin of a particle can be calculated from the number of up and down quarks it contains.

kaon A group of spin-0 mesons consisting of up, down, and strange quarks and their anti-quarks. These are K^+ (up-anti-strange), K^- (strange-anti-up), K^0 (mixtures of down-anti-strange and strange-anti-down) with masses 493.7 MeV/c^2 ($K^\pm$) and 497.6 MeV/c^2 (K^0).

Lamb shift A small difference between two electron energy levels of the hydrogen atom, discovered by Willis Lamb and Robert Retherford in 1947. The Lamb shift provided an important clue which led to the development of renormalization and ultimately quantum electrodynamics.

Λ-CDM An abbreviation of Lambda-cold dark matter. Also known as the 'concordance model' or the 'standard model' of big bang cosmology. The Λ-CDM model accounts for the large-scale structure of the universe, the cosmic microwave background radiation, the accelerating expansion of the universe, and the distribution of elements such as hydrogen, helium, lithium, and oxygen. The model assumes that 69.1 per cent of the mass-energy of the universe is dark energy (reflected in the size of the cosmological constant, Λ), 26.0 per cent is cold dark matter, leaving the visible universe—galaxies, stars, planets, gas, and dust—to account for just 4.9 per cent.

Leggett inequality Named for English physicist Anthony Leggett as an extension of the logic of Bell's theorem and Bell's inequality. The introduction of local hidden variables implies two logical consequences: measurements involving entangled pairs are not affected by the way the experimental apparatus is *set up* and they are not affected by the *results* of any measurement that is actually made on one, the other, or both particles in the pair. Leggett defined a class of 'crypto' non-local hidden variable theories in which the experimental set up can indeed affect the outcome, but the actual results cannot. Such theories do not predict all the possible results that quantum theory predicts and Leggett was able to develop an inequality that could be subjected to a direct test.

lepton From the Greek *leptos,* meaning small. Leptons form a class of particles which do not experience the strong nuclear force and combine with quarks to form matter. Like quarks, leptons form three generations, including the electron, muon, and tau with electric charge −1, spin ½ and masses 0.51 MeV/c^2, 105.7 MeV/c^2, and 1.78 GeV/c^2, respectively, and their corresponding neutrinos. The electron, muon, and tau neutrinos carry no electric charge, have spin ½, and are believed to possess very small masses (necessary to explain the phenomenon of neutrino oscillation, the quantum-mechanical mixing of neutrino flavours such that the flavour may change over time).

LHC Acronym for Large Hadron Collider. The world's highest-energy particle collider, designed to produce proton–proton collision energies of 14 TeV. The LHC is 27 kilometres in circumference and lies 175 metres beneath the Swiss–French border at CERN, near Geneva. The LHC, operating at proton–proton collision energies of 7 and 8 TeV, produced evidence which led to the discovery of the Higgs boson in July 2012. After a two-year shutdown, it began operations in 2015 at a collision energy of 13 TeV.

mass In classical mechanics, the mass of a physical object is a measure of its resistance to changes in its state of motion under the influence of an applied force, assumed to be related to the 'quantity of matter' that it contains. As such, it is a 'primary' quality of material substance. In special relativity and quantum physics, our understanding of the nature of mass changes quite dramatically. Mass becomes a measure of the energy content of an object ($m = E/c^2$), and the mass of elementary particles is traced to the energies associated with different kinds of quantum fields.

mass renormalization *See* renormalization.

mechanical philosophy A branch of seventeenth-century natural philosophy concerned with the establishment of a particularly mechanistic view of nature. The approach of the mechanical philosophers was determinedly reductionist—the properties and behaviour of all natural objects (including living creatures) can be understood in terms of mechanical principles. The mechanical philosophers ushered in the modern scientific era, and included: Francis Bacon (born 1561), Galileo Galilei (1564), Johannes Kepler (1571), Pierre Gassendi (1592), Robert Boyle (1627), Christian Huygens (1629), and Isaac Newton (1642).

mega A prefix denoting million. A mega electron volts (MeV) is a million electron volts, 10^6 eV.

meson From the Greek *mésos*, meaning 'middle'. Mesons are a sub-class of hadrons. They experience the strong nuclear force and are composed of quarks and anti-quarks.

mole A standard unit for the amount of a chemical substance, defined as the amount which contains as many atoms or molecules of the substance as there are carbon atoms in 12 grams of carbon-12 (6.022×10^{23}). It is approximately equal to the atomic or molecular weight of the substance in grams. The name is derived from 'molecule'. *See also* Avogadro's constant.

molecule A fundamental unit of chemical substance formed from two or more atoms. A molecule of oxygen consists of two oxygen atoms, O_2. A molecule of water consists of two hydrogen atoms and one oxygen atom, H_2O.

muon A second-generation lepton equivalent to the electron, with a charge -1, spin $\frac{1}{2}$ (fermion), and mass 105.7 MeV/c^2. First discovered in 1936 by Carl Anderson and Seth Neddermeyer.

Nambu–Goldstone boson Massless, spin-0 particles created as a consequence of spontaneous symmetry-breaking, first discovered by Yoichiro Nambu in 1960 and elaborated by Jeffrey Goldstone in 1961.

natural minima, or *minima naturalia* The ancient Greek philosopher Aristotle hypothesized that there must be 'smallest parts'—natural minima—into which a naturally occurring substance could be divided without losing its essential character. Such smallest parts are *not* atoms. In Aristotle's philosophy they would still be imbued with the form of the substance and can in principle be further divided. However, any further division results in a loss of form—what remains can no longer be considered to be the original substance.

neutrino From Italian, meaning 'small neutral one'. Neutrinos are the chargeless, spin ½ (fermion) companions to the negatively charged electron, muon, and tau. The neutrinos are believed to possess very small masses, necessary to explain the phenomenon of neutrino oscillation, the quantum-mechanical mixing of neutrino flavours such that the flavour may change over time. Neutrino oscillation solves the solar neutrino problem—the problem that the numbers of neutrinos measured to pass through the Earth are inconsistent with the numbers of electron neutrinos expected from nuclear reactions occurring in the Sun's core. It was determined in 2001 that only thirty-five per cent of the neutrinos from the Sun are electron neutrinos—the balance are muon and tau neutrinos, indicating that the neutrino flavours oscillate as they travel from the Sun to the Earth.

neutron An electrically neutral sub-atomic particle, first discovered in 1932 by James Chadwick. The neutron is a baryon consisting of one-up and two-down quarks with spin ½ and mass 939.6 MeV/c^2.

Noether's theorem Developed by Amalie Emmy Noether in 1918. The theorem connects the laws of conservation with specific continuous symmetries of physical systems and the theories that describe them, used as a tool in the development of new theories. The conservation of energy reflects the fact that the laws governing energy are symmetric to continuous changes in time. For linear momentum, the laws are symmetric to continuous changes in spatial position. For angular momentum, the laws are symmetric to the *angle* of direction measured from the centre of the rotation.

nucleus The dense region at the core of an atom in which most of the atom's mass is concentrated. Atomic nuclei consist of varying numbers of protons and neutrons. The nucleus of a hydrogen atom consists of a single proton.

Pauli exclusion principle Discovered by Wolfgang Pauli in 1925. The exclusion principle states that no two fermions may occupy the same quantum state (i.e., possess the same set of quantum numbers) simultaneously. For electrons in atoms, this means that only two electrons can occupy a single atomic orbital provided that they possess opposite spins.

perihelion If a planet were to describe a circular orbit around the Sun, then there would obviously be no change in the distance between the

Sun and the planet as it moves around its orbit. However, the planets of the solar system describe elliptical orbits with the Sun at one focus. This means that the distance between the Sun and the planet does change. The perihelion is the point in the orbit at which the planet is closest to the Sun. The aphelion is the point at which the planet is furthest from the Sun. At its perihelion, the Earth is about 147.1 million kilometres from the Sun. At aphelion it is 152.1 million kilometres from the Sun.

perturbation theory A mathematical method used to find approximate solutions to equations that cannot be solved exactly. The offending equation is recast as a perturbation expansion: the sum of a potentially infinite series of terms which starts with a 'zeroth-order' expression involving no interaction which can be solved exactly. To this is added additional interaction (or perturbation) terms representing corrections to first-order, second-order, third-order, etc. In principle, each term in the expansion provides a smaller and smaller correction to the zeroth-order result, gradually bringing the calculation closer and closer to the actual result. The accuracy of the final result then depends simply on the number of perturbation terms included in the calculation.

photon The elementary particle underlying all forms of electromagnetic radiation, including light. The photon is a massless, spin-1 boson, which acts as the carrier of the electromagnetic force.

pion A group of spin-0 mesons formed from up and down quarks and their anti-quarks. These are π^+ (up-anti-down), π^- (down-anti-up) and π^0 (a mixture of up-anti-up and down-anti-down), with masses 139.6 MeV/c^2 ($\pi^\pm$) and 135.0 MeV/c^2 (π^0). The pions can be thought of as the 'carriers' of the strong force binding protons and neutrons inside the atomic nucleus, representing a kind of 'leakage' of the colour force binding quarks inside the protons and neutrons beyond the boundaries of these particles. First hypothesized by Japanese physicst Hideki Yukawa in 1935.

Planck constant Denoted h. Discovered by Max Planck in 1900. The Planck constant is a fundamental physical constant which reflects the magnitudes of quanta in quantum theory. For example, the energies of photons are determined by their radiation frequencies according to the relation $E = h\nu$, energy equals Planck's constant multiplied by the radiation frequency. Planck's constant has the value 6.626×10^{-34} Joule-seconds.

Planck length The ultimate unit of length, derived from a collection of fundamental physical constants as $\sqrt{(hG/2\pi c^3)}$, where h is Planck's constant, G is Newton's gravitational constant and c is the speed of light. It has the value 1.616×10^{-35} m. The Planck length helps to define the Planck scale, along with the Planck mass $\sqrt{(hc/2\pi G)}$, about 2.177×10^{-8} kg, and the Planck time $\sqrt{(hG/2\pi c^5)}$, the time taken for light to travel the Planck length, 5.391×10^{-44} s.

positron The anti-particle of the electron, denoted e^+, with a charge $+1$, spin $\frac{1}{2}$ (fermion), and mass 0.511 MeV/c^2. The positron was the first anti-particle to be discovered, by Carl Anderson in 1932.

primary quality A concept developed by the classical modern and mechanical philosophers in the seventeenth century, most closely associated with English philosopher John Locke. The primary qualities of an object are independent of observation and include things like solidity, extension (in three-dimensional space), motion, number, and so on. *See also* Secondary quality.

proton A positively charged sub-atomic particle 'discovered' and so named by Ernest Rutherford in 1919. Rutherford actually identified that the nucleus of the hydrogen atom (which is a single proton) is a fundamental constituent of other atomic nuclei. The proton is a baryon consisting of two-up and one-down quarks with spin $\frac{1}{2}$ and mass 938.3 MeV/c^2.

quantum A fundamental, indivisible unit of properties such as energy and angular momentum. In quantum theory, such properties are recognized not to be continuously variable but to be organized in discrete packets or bundles, called quanta. In quantum field theory the use of the term is extended to include particles. Thus, the photon is the quantum particle of the electromagnetic field. This idea can be extended beyond the carriers of forces to include matter particles themselves. Thus, the electron is the quantum of the electron field, and so on. This is sometimes referred to as second quantization.

quantum chromodynamics (QCD) The quantum field theory of the colour force between quarks carried by a system of eight coloured gluons.

quantum electrodynamics (QED) The quantum field theory of the electromagnetic force between electrically charged particles, carried by photons.

equations, whilst serving in the German Army in 1916. The Schwarzschild solutions establish a fundamental boundary, called the *Schwarzschild radius*. A spherical mass, *m*, compressed to a radius less than the Schwarzschild radius (given by Gm/c^2, where *c* is the speed of light and *G* is the gravitational constant) will become a black hole—its escape velocity exceeds the speed of light.

secondary quality A concept developed by the classical modern and mechanical philosophers in the seventeenth century, most closely associated with English philosopher John Locke. The secondary qualities of an object result from the sensations that it produces in the mind of an observer, and include things like colour, taste, touch, sound, and smell. *See also* Primary quality.

self-energy In a quantum field theory, particles are envisaged as fundamental fluctuations or vibrations of the field. One consequence is that particles may undergo self-interaction—they interact with their own fields. Such interactions increase the energy of the particle by an amount called self-energy. In early versions of the quantum field theory of electrons, the self-energy was found to be infinite. This problem was resolved by applying the techniques of renormalization.

spacetime and spacetime metric The distance between one position in a co-ordinate system and another can be determined from the values of the co-ordinates at these two positions. So, in a three-dimensional Euclidean space, if the positions are $l_1 = x_1 y_1 z_1$ and $l_2 = x_2 y_2 z_2$, the distance $\Delta l = l_2 - l_1$ can be found by applying Pythagoras' theorem: $\Delta l^2 = \Delta x^2 + \Delta y^2 + \Delta z^2$. This 'distance function' is often referred to as a *metric*. It has an important property: no matter how we define the co-ordinate system (no matter how we define *x*, *y*, and *z*), the metric will always be the same (mathematicians say that it is 'invariant'). We can extend Euclidean space to include a fourth dimension of time. To ensure that the resulting spacetime metric is invariant we need a structure such as $\Delta s^2 = \Delta(ct)^2 - \Delta x^2 - \Delta y^2 - \Delta z^2$, where *s* is a generalized spacetime interval, *t* is time, and *c* is the speed of light. We could swop these around and define Δs^2 such that the time interval is negative and the spatial intervals positive—so long as these are of opposite sign, Δs^2 is invariant. The choice of signs is a simply a matter of convention.

special relativity Developed by Einstein in 1905, the special theory of relativity asserts that all motion is relative, and there is no unique or privileged frame of reference against which motion can be measured. All inertial frames of reference are equivalent—an observer stationary on Earth should obtain the same results from the same set of physical measurements as an observer moving with uniform velocity in a spaceship. Out go classical notions of absolute space, time, absolute rest, and simultaneity. In formulating the theory, Einstein assumed that the speed of light in a vacuum represents an ultimate speed, which cannot be exceeded. The theory is 'special' only in the sense that it does not account for accelerated motion or gravity; this is covered in Einstein's general theory of relativity.

spectrum Any physical property which has a range of possible values may be said to have a 'spectrum'. The most obvious example is the range of colours produced when light is passed through a prism or a collection of raindrops to produce a rainbow. The resulting spectrum may appear continuous (as in a rainbow) or it may be discrete, consisting of a set of specific values. The absorption or emission spectrum of hydrogen atoms exhibits a series of 'lines' corresponding to radiation frequencies that are absorbed or emitted by the atoms. The positions (frequencies) of these lines in the spectrum relate to the energies of the electron orbitals involved.

spin All elementary particles exhibit a type of angular momentum called spin. Although the spin of the electron was initially interpreted in terms of 'self-rotation' (the electron spinning on its own axis, like a spinning top), spin is a relativistic phenomenon and has no counterpart in classical physics. Particles are characterized by their spin quantum numbers. Particles with half-integral spin quantum numbers are called fermions. Particles with integral spin quantum numbers are called bosons. Matter particles are fermions. Force particles are bosons.

standard model, of big bang cosmology *See* Λ-CDM.

standard model, of particle physics The currently-accepted theoretical model describing matter particles and the forces between them, with the exception of gravity. The standard model consists of a collection of quantum field theories which describes three generations of quarks and leptons, the photon, W, and Z particles, colour force gluons, and the Higgs boson.

standing wave Waves confined to oscillate between fixed points will interfere and may settle into a series of standing or stationary wave patterns. This is the basis for the production of musical notes in string or wind instruments, and there are many examples of naturally occurring standing waves in the atmosphere close to mountains and in river rapids. A condition for establishing standing waves in a stationary medium is that the waves must contain an integral number of half-wavelengths.

strange quark A second-generation quark with charge $-\frac{1}{3}$, spin $\frac{1}{2}$ (fermion), and a mass of 95 MeV/c^2. The property of 'strangeness' was identified as a characteristic of a series of relatively low energy (low mass) particles discovered in the 1940s and 1950s by Murray Gell-Mann and independently by Kazuhiko Nishijima and Tadao Nakano. This property was subsequently traced by Gell-Mann and George Zweig to the presence in these composite particles of the strange quark.

strangeness Identified as a characteristic property of particles such as the neutral lambda, neutral and charged sigma and xi particles, and the kaons. Strangeness was used together with electric charge and isospin to classify particles according to the 'Eightfold Way' by Murray Gell-Mann and Yuval Ne'eman. This property was subsequently traced to the presence in these composite particles of the strange quark.

strong force The strong nuclear force, or colour force, binds quarks and gluons together inside hadrons and is described by quantum chromodynamics. The force that binds protons and neutrons together inside atomic nuclei (also referred to as the strong nuclear force) is thought to be the result of a 'leakage' of the colour force binding quarks inside the nucleons. *See* Colour force.

SU(2) symmetry group The special unitary group of transformations of two complex variables. Identified by Chen Ning Yang and Robert Mills as the symmetry group on which a quantum field theory of the strong nuclear force should be based, SU(2) was subsequently identified with the weak force and can also be used to describe the residual strong force acting between protons and neutrons, carried by pions.

SU(3) symmetry group The special unitary group of transformation of three complex variables. Used by Gell-Mann, Harald Fritzsch, and Heinrich Leutwyler as a local symmetry on which to base a quantum field theory of the colour force between quarks and gluons.

substantial form The theory of forms originated with the ancient Greek philosopher Plato but it was his student, Aristotle, who most clearly distinguished between matter (the undifferentiated 'stuff' of material substance) and form (which differentiates and organizes matter and gives an object its essential characteristics). In medieval philosophy, the notion of substantial forms was used to rationalize many aspects of Christian doctrine, such as transubstantiation.

superconductivity Discovered by Heike Kamerlingh Onnes in 1911. When cooled below a certain critical temperature, certain crystalline materials lose all electrical resistance and become superconductors. An electric current will flow indefinitely in a superconducting wire flowing with no energy input. Superconductivity is a quantum mechanical phenomenon explained using the BCS mechanism, named for John Bardeen, Leon Copper, and John Schrieffer.

superposition In quantum mechanics, quantum entities can behave like particles and they can also behave like waves. But waves can be combined—they can be added together in a 'superposition'. Such combinations describe diffraction and interference effects. In a quantum measurement, it is necessary to form a superposition which contains contributions from the wavefunctions describing all the different possible outcomes. The square of the amplitude of each wavefunction in the superposition relates to the probability that the corresponding outcome will be observed. When the measurement is made, the wavefunction 'collapses' and all the other possible outcomes disappear.

symmetry-breaking Spontaneous symmetry-breaking occurs whenever the lowest energy state of a physical system has lower symmetry than higher-energy states. As the system loses energy and settles to its lowest energy state, the symmetry spontaneously reduces, or 'breaks'. For example, a pencil perfectly balanced on its tip is symmetrical, but under the influence of the background environment (such as small currents or air) it will topple over to give a more stable, lower-energy, less symmetrical state, with the pencil lying along one specific direction.

tera A prefix denoting trillion. A tera electron volts (TeV) is a trillion electron volts, 10^{12} eV, or 1,000 GeV.

top quark Also sometimes referred to as the 'truth' quark. A third-generation quark with charge $+\frac{2}{3}$, spin $\frac{1}{2}$ (fermion), and a mass of 173 GeV/c^2. It was discovered at Fermilab in 1995.

trillion A thousand billion or a million million, 10^{12}, or 1,000,000,000,000.

U(1) symmetry group The unitary group of transformations of one complex variable. It is equivalent (the technical term is 'isomorphic') with the circle group, the multiplicative group of all complex numbers with absolute value of unity (in other words, the unit circle in the complex plane). It is also isomorphic with SO(2), a special orthogonal group which describes the symmetry transformation involved in rotating an object in two dimensions. In quantum electrodynamics, U(1) is identified with the phase symmetry of the electron wavefunction.

uncertainty principle Discovered by Werner Heisenberg in 1927. The uncertainty principle states that there is a fundamental limit to the precision with which it is possible to measure pairs of 'conjugate' observables, such as position and momentum and energy and the rate of change of energy with time. The principle can be traced to the fundamental duality of wave and particle behaviour in quantum objects.

void In the atomic theory of the ancient Greek philosophers, matter was thought to consist of tiny, indivisible atoms moving restlessly in empty space, which was called the void. In this context, void simply means space empty or devoid of any matter, and which today we tend to call a vacuum.

W, Z particles Elementary particles which carry the weak nuclear force. The W particles are spin-1 bosons with unit positive and negative electrical charge (W^+, W^-) and masses of 80.4 GeV/c^2. The Z^0 is an electrically neutral spin-1 boson with mass 91.2 GeV/c^2. The W and Z particles gain mass through the Higgs mechanism.

wave–particle duality A fundamental property of all quantum particles, which exhibit both delocalized wave behaviour (such as diffraction and interference) and localized particle behaviour, depending on the type of apparatus used to make measurements on them. First suggested as a property of matter particles such as electrons by Louis de Broglie in 1923.

wavefunction The mathematical description of matter particles such as electrons as 'matter waves' leads to equations characteristic of wave motion. Such wave equations feature a wavefunction whose amplitude and phase evolves in space and time. The wavefunctions of the electron in a hydrogen atom form characteristic three-dimensional patterns around the nucleus called orbitals. Wave mechanics—an expression of quantum mechanics in terms of matter waves—was first elucidated by Erwin Schrödinger in 1926.

wavefunction collapse *See* Collapse of the wavefunction.

weak force The weak force is so called because it is considerably weaker than both the strong and electromagnetic forces, in strength and range. The weak force affects both quarks and leptons and weak force interactions can change quark and lepton flavour: for example, turning an up quark into a down quark and an electron into an electron neutrino. The weak force was originally identified as a fundamental force from studies of beta radioactive decay. Carriers of the weak force are the W and Z particles. The weak force was combined with electromagnetism in the quantum field theory of the electro-weak force by Steven Weinberg and Abdus Salam in 1967–1968.

Yang–Mills field theory A form of quantum field theory developed in 1954 by Chen Ning Yang and Robert Mills. Yang–Mills field theories underpin all the components of the current standard model of particle physics.

SELECT BIBLIOGRAPHY

It goes without saying that the development of our understanding of the nature of matter has a long and illustrious history, and enough has been written on the subject to fill whole libraries. The following is a selection of the books that I have found most helpful not only in writing *Mass* but also in helping to clarify the historical development and many of the basic principles of the science involved. I heartily recommend them to readers in search of further enlightenment. Titles marked with an asterisk are more suitable for readers with a scientific background.

ANTHOLOGIES

FRENCH, A.P. and KENNEDY, P.J. (eds), *Niels Bohr: A Centenary Volume*, Harvard University Press, Cambridge, MA, 1985.

SCHILPP, PAUL ARTHUR (ed.), *Albert Einstein. Philosopher-scientist*, The Library of Living Philosophers, Vol. 1, Harper & Row, New York, 1959 (first published 1949).

*WHEELER, JOHN ARCHIBALD and ZUREK, WOJCIECH HUBERT (eds), *Quantum Theory and Measurement*, Princeton University Press, 1983.

BIOGRAPHIES

BERNSTEIN, JEREMY, *Quantum Profiles*, Princeton University Press, 1991.

CASSIDY, DAVID C., *Uncertainty: The Life and Science of Werner Heisenberg*, W.H. Freeman, New York, 1992.

DYSON, FREEMAN, *Disturbing the Universe*, Basic Books, New York, 1979.

*ENZ, CHARLES P., *No Time to be Brief: A Scientific Biography of Wolfgang Pauli*, Oxford University Press, 2002.

FARMELO, GRAHAM, *The Strangest Man: The Hidden Life of Paul Dirac, Quantum Genius*, Faber and Faber, London, 2009.

FEYNMAN, RICHARD P., *'Surely You're Joking, Mr. Feynman!'*, Unwin, London, 1985.

GLEICK, JAMES, *Genius: Richard Feynman and Modern Physics*, Little, Brown & Co., London, 1992.

GLEICK, JAMES, *Isaac Newton*, Harper Perennial, London, 2004.

HEILBRON, J.L., *The Dilemmas of an Upright Man: Max Planck and the Fortunes of German Science*, Harvard University Press, Cambridge, MA, 1996.

ISAACSON, WALTER, *Einstein: His Life and Universe*, Simon & Shuster, New York, 2007.

JOHNSON, GEORGE, *Strange Beauty: Murray Gell-Mann and the Revolution in Twentieth-century Physics*, Vintage, London, 2001.

*KRAGH, HELGE, *Dirac: A Scientific Biography*, Cambridge University Press, 1990.

MEHRA, JAGDISH, *The Beat of a Different Drum: The Life and Science of Richard Feynman*, Oxford University Press, 1994.

MOORE, WALTER, *Schrödinger: Life and Thought*, Cambridge University Press, 1989.

*PAIS, ABRAHAM, *Subtle is the Lord: The Science and the Life of Albert Einstein*, Oxford University Press, 1982.

*PAIS, ABRAHAM, *Niels Bohr's Times, in Physics, Philosophy and Polity*, Clarendon Press, Oxford, 1991.

PEAT, F. DAVID, *Infinite Potential: The Life and Times of David Bohm*, Addison-Wesley, Reading, MA, 1997.

WHEELER, JOHN ARCHIBALD, with FORD, KENNETH, *Geons, Black Holes and Quantum Foam: A Life in Physics*, W.W. Norton & Company, Inc., New York, 1998.

COSMOLOGY

GOLDSMITH, DONALD, *The Runaway Universe: The Race to Find the Future of the Cosmos*, Perseus Publishing, New York, 2000.

GUTH, ALAN H., *The Inflationary Universe: The Quest for a New Theory of Cosmic Origins*, Vintage, London, 1998.

HAWKING, STEPHEN, *A Brief History of Time: From the Big Bang to Black Holes*, Bantam Press, London, 1988.

KIRSHNER, ROBERT P., *The Extravagant Universe: Exploding Stars, Dark Energy and the Accelerating Cosmos*, Princeton University Press, 2002.

KRAGH, HELGE, *Masters of the Universe: Conversations with Cosmologists of the Past*, Oxford University Press, 2015.

KRAUSS, LAWRENCE M., *A Universe from Nothing: Why There is Something Rather than Nothing*, Simon & Schuster, London, 2012.

*NUSSABAUMER, HARRY and BIERI, LYDIA, *Discovering the Expanding Universe*, Cambridge University Press, 2009.

OSTRIKER, JEREMIAH P. and MITTON, SIMON, *Heart of Darkness: Unravelling the Mysteries of the Invisible Universe*, Princeton University Press, 2013.

OVERBYE, DENNIS, *Lonely Hearts of the Cosmos: The Quest for the Secret of the Universe*, Picador, London, 1993.

PANEK, RICHARD, *The 4% Universe: Dark Matter, Dark Energy and the Race to Discover the Rest of Reality*, Oneworld, Oxford, 2011.

REES, MARTIN, *Just Six Numbers: The Deep Forces that Shape the Universe*, Phoenix, London, 2000.

SINGH, SIMON, *Big Bang: The Most Important Scientific Discovery of All Time and Why You Need to Know About It*, Harper Perennial, London, 2005.

WEINBERG, STEVEN, *The First Three Minutes: A Modern View of the Origin of the Universe*, Basic Books, New York, 1977.

*WEINBERG, STEVEN, *Cosmology*, Oxford University Press, 2008.

GENERAL

CREASE, ROBERT P., *A Brief Guide to the Great Equations: The Hunt for Cosmic Beauty in Numbers*, Robinson, London, 2009.

FARMELO, GRAHAM (ed.), *It Must Be Beautiful: Great Equations of Modern Science*, Granta Books, London, 2002.

FEYNMAN, RICHARD, *The Character of Physical Law*, MIT Press, Cambridge, MA, 1967.

*FEYNMAN, RICHARD P., LEIGHTON, ROBERT B., and SANDS, MATTHEW, *The Feynman Lectures on Physics*, Vol. III, Addison-Wesley, Reading, MA, 1965.

GREENBLATT, STEPHEN, *The Swerve: How the Renaissance Began*, Vintage, London, 2012.

*KENNEDY, ROBERT E., *A Student's Guide to Einstein's Major Papers*, Oxford University Press, 2012.

KRAGH, HELGE, *Quantum Generations: A History of Physics in the Twentieth Century*, Princeton University Press, 1999.

ROVELLI, CARLO, *Seven Brief Lessons on Physics*, Allen Lane, London, 2015.

*STACHEL, JOHN (ed.), *Einstein's Miraculous Year: Five Papers that Changed the Face of Physics*, Princeton University Press, 2005.

HISTORY AND PHILOSOPHY

ARISTOTLE, *The Metaphysics*, trans. John H. McMahon, Prometheus Books, New York, 1991.

*BUCHWALD, JED Z., *The Rise of the Wave Theory of Light*, University of Chicago Press, 1989.

CARTWRIGHT, NANCY, *How the Laws of Physics Lie*, Oxford University Press, 1983.

CHALMERS, A.F., *What is This Thing Called Science?*, 3rd edn, Hackett, IN, 1999.

*CHALMERS, ALAN, *The Scientist's Atom and the Philosopher's Stone: How Science Suceeded and Philosophy Failed to Gain Knowledge of Atoms*, Springer, London, 2011.

CLARK, MICHAEL, *Paradoxes from A to Z.* Routledge, London, 2002.

CUSHING, JAMES T., *Philosophical Concepts in Physics*, Cambridge University Press, 1998.

D'ESPAGNAT, BERNARD, *Reality and the Physicist*, Cambridge University Press, 1989.

DESCARTES, RENE, *Discourse on Method and the Meditations*, trans. F.E. Sutcliffe, Penguin, London, 1968.

FARA, PATRICIA, *Science: A Four Thousand Year History*, Oxford University Press, 2009.

FEYERABEND, PAUL, *Farewell to Reason*, Verso, London, 1987.

FEYERABEND, PAUL, *Against Method*, 3rd edn, Verso, London, 1993.

GARDNER, SEBASTIAN, *Kant and the Critique of Pure Reason*, Routledge, Abingdon, 1999.

GILLIES, DONALD, *Philosophy of Science in the Twentieth Century*, Blackwell, Oxford, 1993.

GREENAWAY, FRANK, *John Dalton and the Atom*, Heinemann, London, 1966.

HACKING, IAN, *Representing and Intervening*, Cambridge University Press, 1983.

*HERMANN, ARMIN, *The Genesis of Quantum Theory (1899–1913)*, trans. Claude W. Nash, MIT Press, Cambridge, MA, 1971.

HUME, DAVID, *A Treatise of Human Nature*, Penguin, London, 1969.

*JAMMER, MAX, *The Philosophy of Quantum Mechanics: The Interpretations of Quantum Mechanics in Historical Perspective*, John Wiley & Sons, Inc., New York, 1974.

*JAMMER, MAX, *Concepts of Mass in Contemporary Physics and Philosophy*, Princeton University Press, 2000.

JARDINE, LISA, *The Curious Life of Robert Hooke: The Man Who Measured London*, Harper Collins, London, 2003.

KANT, IMMANUEL, *Critique of Pure Reason*, trans. J.M.D. Meiklejohn, J.M. Dent & Sons, London, 1988.

KUHN, THOMAS S., *The Structure of Scientific Revolutions*, 2nd edn, University of Chicago Press, 1970.

LAËRTIUS, DIOGENES, *Lives of the Eminent Philosophers*, trans. Robert Drew Hicks, Loeb Classical Library (1925), http://en.wikisource.org/wiki/Lives_of_the_Eminent_Philosophers.

LEIBNIZ, GOTTFRIED WILHELM, *Philosophical Writings*, trans. Mary Morris and G.H.R. Parkinson, J.M. Dent & Sons, London, 1973.

LUCRETIUS, *On the Nature of the Universe*, trans. R.E. Latham, Penguin Books, London (first published 1951).

NOLA, ROBERT and SANKEY, HOWARD, *Theories of Scientific Method*, Acumen, Durham, 2007.

PLATO, *Timaeus and Critias*, Penguin, London. 1971.

POPPER, KARL R., *The Logic of Scientific Discovery*, Hutchinson, London, 1959.

PSILLOS, STATHIS, *Scientific Realism: How Science Tracks Truth*, Routledge, London, 1999.

PSILLOS, STATHIS and CURD, MARTIN (eds), *The Routledge Companion to Philosophy of Science*, Routledge, London, 2010.

SAMBURSKY, SAMUEL, *The Physical World of the Greeks*, 2nd edn, Routledge, London, 1960.

SCHACHT, RICHARD, *Classical Modern Philosophers: Descartes to Kant*, Routledge and Kegan Paul, London, 1984.

*TOMONAGA, SIN-ITIRO, *The Story of Spin*, University of Chicago Press, 1997.

VAN FRAASEN, BAS C., The Scientific Image, Oxford University Press, 1980.

WILSON, CATHERINE, Epicureanism at the Origins of Modernity, Oxford University Press, 2008.

QUANTUM THEORY

ACZEL, AMIR D., Entanglement: The Greatest Mystery in Physics, John Wiley & Sons Ltd, Chichester, 2003.

BAGGOTT, JIM, Beyond Measure: Modern Physics, Philosophy and the Meaning of Quantum Theory, Oxford University Press, 2004.

BAGGOTT, JIM, The Quantum Story: A History in 40 Moments, Oxford University Press, 2011.

*BELL, J.S., Speakable and Unspeakable in Quantum Mechanics, Cambridge University Press, 1987.

BELLER, MARA, Quantum Dialogue. University of Chicago Press, 1999.

BOHN, DAVID, Causality and Chance in Modern Physics, Routledge & Kegan Paul, London, 1984.

DAVIES, P.C.W. and BROWN, J.R. (eds), The Ghost in the Atom, Cambridge University Press, 1986.

*D'ESPAGNAT, BERNARD, Conceptual Foundations of Quantum Mechanics, 2nd edn, Addison-Wesley, Reading, MA, 1976.

FEYNMAN, RICHARD P., QED: The Strange Theory of Light and Matter, Penguin, London, 1985.

FINE, ARTHUR, The Shaky Game: Einstein, Realism and the Quantum Theory, 2nd edn, University of Chicago Press, 1986.

*FRENCH, A.P. and TAYLOR, E.F., An Introduction to Quantum Physics, Van Nostrand Reinhold, Wokingham, 1978.

GAMOW, GEORGE, Thirty Years that Shook Physics, Dover Publications, New York, 1966.

GRIBBIN, JOHN, Schrödinger's Kittens, Phoenix, London, 1996.

HEISENBERG, WERNER, The Physical Principles of the Quantum Theory, trans. Carl Eckart and F.C. Hoyt, Dover, New York, 1949.

HEISENBERG, WERNER, Physics and Philosophy: The Revolution in Modern Science, Penguin, London, 1989 (first published 1958).

*KUHN, THOMAS, S., Black-Body Theory and the Quantum Discontinuity 1894–1912, Oxford University Press, 1978.

KUMAR, MANJIT, *Quantum: Einstein, Bohr and the Great Debate About the Nature of Reality*, Icon Books, London, 2008.

LINDLEY, DAVID, *Where Does the Weirdness Go? Why Quantum Mechanics is Strange, but Not as Strange as You Think*, BasicBooks, New York, 1996.

MEHRA, JAGDISH, *Einstein, Physics and Reality*, World Scientific, London, 1999.

ORZEL, CHAD, *How to Teach Quantum Physics to Your Dog*, Oneworld, London, 2010.

POPPER, KARL R., *Quantum Theory and the Schism in Physics*, Unwin Hyman, London, 1982.

RAE, ALASTAIR, *Quantum Physics: Illusion or Reality?*, Cambridge University Press, 1986.

*RAE, ALASTAIR I.M., *Quantum Mechanics*, 2nd edn, Adam Hilger, Bristol, 1986.

*SCHWEBER, SILVAN S., *QED and the Men Who Made It: Dyson, Feynman, Schwinger, Tomonaga*, Princeton University Press, 1994.

*WAERDEN, B.L. VAN DER, *Sources of Quantum Mechanics*, Dover, New York, 1968.

*ZEE, A., *Quantum Field Theory in a Nutshell*, Princeton University Press, 2003.

PARTICLE PHYSICS

BAGGOTT, JIM, *Higgs: The Invention and Discovery of the 'God Particle'*, Oxford University Press, 2012.

CARROLL, SEAN, *The Particle at the End of the Universe: The Hunt for the Higgs and the Discovery of a New World*, Oneworld, London, 2012.

CLOSE, FRANK, *Antimatter*, Oxford University Press, 2009.

CLOSE, FRANK, *The Infinity Puzzle: How the Quest to Understand Quantum Field Theory Led to Extraordinary Science, High Politics, and the World's Most Expensive Experiment*, Oxford University Press, 2011.

*CREASE, ROBERT P. and MANN, CHARLES C., *The Second Creation: Makers of the Revolution in Twentieth-century Physics*, Rutgers University Press, 1986.

GELL-MANN, MURRAY, *The Quark and the Jaguar*, Little, Brown & Co., London, 1994.

HALPERN, PAUL, *Collider: The Search for the World's Smallest Particles*, John Wiley & Sons, Inc., Somerset, NJ, 2009.

*HODDESON, LILLIAN, BROWN, LAURIE, RIORDAN, MICHAEL, and DRESDEN, MAX, *The Rise of the Standard Model: Particle Physics in the 1960s and 1970s*, Cambridge University Press, 1997.

LEDERMAN, LEON (with Dick Teresi), *The God Particle: If the Universe is the Answer, What is the Question?*, Bantam Press, London, 1993.

NAMBU, YOICHIRO, *Quarks*, World Scientific Publishing, Singapore, 1981.

PAIS, ABRAHAM, *Inward Bound: Of Matter and Forces in the Physical World*, Oxford University Press, 1986.

RIORDAN, MICHAEL, *The Hunting of the Quark: A True Story of Modern Physics*, Simon & Shuster, New York, 1987.

SAMPLE, IAN, *Massive: The Hunt for the God Particle*, Virgin Books, London, 2010.

T' HOOFT, GERARD, *In Search of the Ultimate Building Blocks*, Cambridge University Press, 1997.

VELTMAN, MARTINUS, *Facts and Mysteries in Elementary Particle Physics*, World Scientific, London, 2003.

WEINBERG, STEVEN, *Dreams of a Final Theory: The Search for the Fundamental Laws of Nature*, Vintage, London, 1993.

PUBLISHER'S ACKNOWLEDGEMENTS

We are grateful for permission to include the following copyright material in this book:

Extract from Paul Dirac, 'The Proton', *Nature*, 126, 1930, pp. 605–6, with permission.

Extract from Lucretius, *On the Nature of the Universe*, trans. R.E. Latham, Penguin Books, London, first published 1951, pp. 61–2. © R.E. Latham, 1951. Reproduced by permission of Penguin Books Ltd.

Extract from Sebastian Gardner, *Kant and the Critique of Pure Reason*, Routledge, Abingdon, 1999, p. 205, with permission.

Extract from Albert Einstein, *Annalen der Physik*, **18** (1905), pp. 639–41. Republished with permission of Princeton University Press, from John Stachel (ed.), *Einstein's Miraculous Year: Five Papers that Changed the Face of Physics*, centenary edn, Princeton University Press, 2005; permission conveyed through Copyright Clearance Center, Inc.

Extract republished with permission of Princeton University Press, from Max Jammer, *Concepts of Mass in Contemporary Physics and Philosophy*, Princeton University Press, 2000, p. 61; permission conveyed through Copyright Clearance Center, Inc.

Extract taken from *GEONS, BLACK HOLES, AND QUANTUM FOAM: A LIFE IN PHYSICS* by John Archibald Wheeler with Kenneth Ford. Copyright © 1998 by John Archibald Wheeler and Kenneth Ford. Used by permission of W.W. Norton & Company, Inc.

Extract from Felix Bloch, 'Heisenberg and the Early Days of Quantum Mechanics', *Physics Today*, **29**, 1976, p. 23, with the permission of the American Institute of Physics, http://dx.doi.org/10.1063/1.3024633.

Extract taken from Hans Bethe, *Calculating the Lamb Shift*, Web of Stories, http://www.webofstories.com/play/hans.bethe/104;jsessionid=45C0C719DE8CEA2C0899D6A63E281F24.

Excerpt from *THE GOD PARTICLE* Leon Lederman with Dick Teresi. Copyright © 1993 by Leon Lederman and Dick Teresi. Reprinted by permission of Houghton Mifflin Publishing Company. All rights reserved. And from *The God Particle* by *Leon Lederman & Dick Teresi*. Published by *Bantam Press*. Reprinted by permission of The Random House Group.

Reprinted by permission of The Random House Group Limited.

Extract from Willis Lamb, *Nobel Lectures, Physics 1942–1962*, Elsevier, Oxford, 1964. Copyright © Elsevier (1964).

The publisher and author have made every effort to trace and contact all copyright holders before publication. If notified, the publisher will be pleased to rectify any errors or omissions at the earliest opportunity.

INDEX

THE QUANTUM STORY

A history in 40 moments

Jim Baggott

978-0-19-878477-7 | Oxford Landmark
Science | Paperback | £10.99

"Jim Baggott's survey of the history of the emergence of the twentieth century's most enigmatic but successful theory is a delight to read. It is clear, accessible, engaging, informative, and thorough. It illuminates an important, revolutionary era of modern science and the varied personalities behind it."

- Peter Atkins

Almost everything we think we know about the nature of our world comes from one theory of physics. Jim Baggott presents a celebration of this wonderful yet wholly disconcerting theory, with a history told in forty episodes—significant moments of truth or turning points in the theory's development. From its birth in the porcelain furnaces used to study black body radiation in 1900, to the promise of stimulating new quantum phenomena to be revealed by CERN's Large Hadron Collider over a hundred years later, this is the extraordinary story of the quantum world.

PHYSICS

A short history from quintessence to quarks

John L. Heilbron

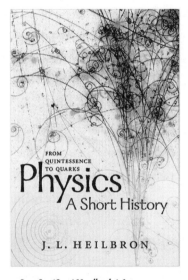

9780198746850 | Hardback | £10.99

"The book is effectively a short history of ideas that moves around the cultures of Europe depending on time and place, so there is a fascinating chapter on Islamic contributions."

- Network Review

"manages to pack an awful lot into that very short space...interesting and informative for non-scientists"

- A Hermit's Progress

How does the physics we know today - a highly professionalised enterprise, inextricably linked to government and industry - link back to its origins as a liberal art in Ancient Greece? What is the path that leads from the old philosophy of nature and its concern with humankind's place in the universe to modern massive international projects that hunt down fundamental particles and industrial laboratories that manufacture marvels?

John Heilbron's fascinating history of physics introduces us to Islamic astronomers and mathematicians, calculating the size of the earth whilst their caliphs conquered much of it; to medieval scholar-theologians investigating light; to Galileo, Copernicus, Kepler, and Newton, measuring, and trying to explain, the universe. We visit the 'House of Wisdom' in 9th-century Baghdad; Europe's first universities; the courts of the Renaissance; the Scientific Revolution and the academies of the 18th century; the increasingly specialised world of 20th and 21st century science.

NEUTRINO

Frank Close

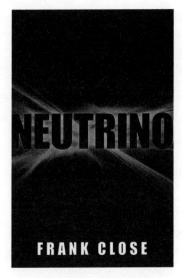

978-0-19-969599-7 | Paperback | £8.99

"Close tells this story with verve and precision...admirably clear and eminently accessible."

- Wall Street Journal

"As an award-winning writer, Close tells this detective story with great style."

- Robert Matthews, BBC Focus

Neutrinos are perhaps the most enigmatic particles in the universe. Formed in certain radioactive decays, they pass through most matter with ease. These tiny, ghostly particles are formed in millions in the Sun and pass through us constantly. For a long time they were thought to be massless, and passing as they do like ghosts they were not regarded as significant. Now we know they have a very small mass, and there are strong indications that they are very important indeed. It is speculated that a heavy form of neutrino, that is both matter and antimatter, may have shaped the balance of matter and antimatter in the early universe.

Here, Frank Close gives an account of the discovery of neutrinos and our growing understanding of their significance, and also touches on some speculative ideas concerning the possible uses of neutrinos and their role in the early universe.

ORIGINS

The Scientific Story of Creation

Jim Baggott

978-0-19-870764-6 | Hardback | £25.00

"The collective mind of humanity has made extraordinary progress in its quest to understand how the current richness of the physical world has emerged, and Baggott with his characteristic lucidity and erudition, has provided an enthralling account of this wonderful and still unfolding intellectual journey."

- Peter Atkins

'There are many different versions of our creation story. This book tells the version according to modern science', writes Jim Baggott. In *Origins*, he presents a unique version of the story in chronological sequence, from the Big Bang to the emergence of human consciousness 13.8 billion years later.

Cosmology, particle physics, chemistry, planetary geology, biology – it is all here, explained with clarity, in one overarching narrative. And throughout, Baggott emphasizes that the scientific story is a work-in-progress, highlighting the many puzzles and uncertainties that still remain. We have a seemingly innate desire to comprehend our own place in the Universe. Jim Baggott helps us fulfil this desire, which is driven in part by simple curiosity but also by a deeper emotional need to connect ourselves meaningfully with the world which we call home.